Recent Titles in This Series

117 **Boris Zilber,** Uncountably categorical theories, 1993
116 **G. M. Fel′dman,** Arithmetic of probability distributions, and characterization problems on abelian groups, 1993
115 **Nikolai V. Ivanov,** Subgroups of Teichmüller modular groups, 1992
114 **Seizô Itô,** Diffusion equations, 1992
113 **Michail Zhitomirskiĭ,** Typical singularities of differential 1-forms and Pfaffian equations, 1992
112 **S. A. Lomov,** Introduction to the general theory of singular perturbations, 1992
111 **Simon Gindikin,** Tube domains and the Cauchy problem, 1992
110 **B. V. Shabat,** Introduction to complex analysis Part II. Functions of several variables, 1992
109 **Isao Miyadera,** Nonlinear semigroups, 1992
108 **Takeo Yokonuma,** Tensor spaces and exterior algebra, 1992
107 **B. M. Makarov, M. G. Goluzina, A. A. Lodkin, and A. N. Podkorytov,** Selected problems in real analysis, 1992
106 **G.-C. Wen,** Conformal mappings and boundary value problems, 1992
105 **D. R. Yafaev,** Mathematical scattering theory: General theory, 1992
104 **R. L. Dobrushin, R. Kotecký, and S. Shlosman,** Wulff construction: A global shape from local interaction, 1992
103 **A. K. Tsikh,** Multidimensional residues and their applications, 1992
102 **A. M. Il′in,** Matching of asymptotic expansions of solutions of boundary value problems, 1992
101 **Zhang Zhi-fen, Ding Tong-ren, Huang Wen-zao, and Dong Zhen-xi,** Qualitative theory of differential equations, 1992
100 **V. L. Popov,** Groups, generators, syzygies, and orbits in invariant theory, 1992
99 **Norio Shimakura,** Partial differential operators of elliptic type, 1992
98 **V. A. Vassiliev,** Complements of discriminants of smooth maps: Topology and applications, 1992
97 **Itiro Tamura,** Topology of foliations: An introduction, 1992
96 **A. I. Markushevich,** Introduction to the classical theory of Abelian functions, 1992
95 **Guangchang Dong,** Nonlinear partial differential equations of second order, 1991
94 **Yu. S. Il′yashenko,** Finiteness theorems for limit cycles, 1991
93 **A. T. Fomenko and A. A. Tuzhilin,** Elements of the geometry and topology of minimal surfaces in three-dimensional space, 1991
92 **E. M. Nikishin and V. N. Sorokin,** Rational approximations and orthogonality, 1991
91 **Mamoru Mimura and Hirosi Toda,** Topology of Lie groups, I and II, 1991
90 **S. L. Sobolev,** Some applications of functional analysis in mathematical physics, third edition, 1991
89 **Valeriĭ V. Kozlov and Dmitriĭ V. Treshchëv,** Billiards: A genetic introduction to the dynamics of systems with impacts, 1991
88 **A. G. Khovanskiĭ,** Fewnomials, 1991
87 **Aleksandr Robertovich Kemer,** Ideals of identities of associative algebras, 1991
86 **V. M. Kadets and M. I. Kadets,** Rearrangements of series in Banach spaces, 1991
85 **Mikio Ise and Masaru Takeuchi,** Lie groups I, II, 1991
84 **Ðáo Trọng Thi and A. T. Fomenko,** Minimal surfaces, stratified multivarifolds, and the Plateau problem, 1991
83 **N. I. Portenko,** Generalized diffusion processes, 1990

(See the AMS catalog for earlier titles)

Translations of
MATHEMATICAL
MONOGRAPHS

Volume 117

Uncountably Categorical Theories

Boris Zilber

American Mathematical Society
Providence, Rhode Island

Translated from the Russian by D. Louvish
Translation edited by Simeon Ivanov

1991 *Mathematics Subject Classification*. Primary 03–02, 03C35, 03C60, 05B25.

Library of Congress Cataloging-in-Publication Data

Zilber, Boris.
[Neschetno kategorichnye teorii. English]
Uncountably categorical theories/Boris Zilber.
p. cm.—(Translations of mathematical monographs, ISSN 0065-9282; v. 117)
ISBN 0-8218-4586-1
1. Categories (Mathematics) 2. Completeness theorem. 3. Model theory. I. Title. II. Series.
QA9.67.Z5513 1993 92-31151
511.3—dc20 CIP

Printed in the United States of America
The paper used in this book is acid-free and falls within the guidelines
established to ensure permanence and durability. ♾
This publication was typeset using AMS-TEX,
the American Mathematical Society's TEX macro system.

10 9 8 7 6 5 4 3 2 1 98 97 96 95 94 93

Contents

Introduction

During the formative period of mathematical logic it was believed that the ideal relationship between a mathematical structure and the formal language is one in which the structure is uniquely determined, up to isomorphism, by its description in the language. Any such description (theory) is commonly called categorical. In the 1930s, however, it became clear from the work of Löwenheim, Skolem, Mal′tsev, and Tarski that for first-order languages (which accord to the greatest degree with the Hilbert program of establishing a finitary foundation for mathematics) the only categorical theories are the theories of finite structures. This entailed a rethinking of the concept of categoricity.

A theory which has, up to isomorphism, a unique model of cardinality λ is said to be λ-categorical. Morley [64], confirming a conjecture of Łos [56], showed that for a countable first-order theory the only possible λ-categorical cases for infinite λ are as follows:

(1) The theory is λ-categorical for all uncountable λ.

(2) The theory is λ-categorical for all infinite λ.

(3) The theory is λ-categorical only for countable λ.

Accordingly, these kinds of theories are known respectively as uncountably, totally, or countably categorical theories. The same terms are used to classify structures which are models of λ-categorical theories. The first class of course contains the second.

As is clear from the definition, the isomorphism type of a λ-categorical structure of cardinality λ is uniquely determined by the cardinality and the first-order axioms. Such structures are interesting both in their own right and because they characterize the expressive capacities of a first-order language.

As the contemporary theory shows, the case of countable categoricity has no deep model-theoretic consequences. The well-known theorem of Ryll-Nardzewski-Svenonius-Engeler is essentially the only general result in this context. In this dissertation our attention will be concentrated on the case of uncountably and, in particular, totally categorical structures.

The intensive study of theories categorical in uncountable powers and of their models, initiated by the ground-breaking paper of Morley [64], has exerted and is still exerting a strong influence on model theory as a whole. In particular, categoricity theory is a basis and source of ideas for stability theory—a technique which is effectively applied in many branches of model theory.

The 1970s saw the appearance and development in categoricity theory of a tendency to focus attention on the study and description of uncountably categorical theories in various special classes defined by natural algebraic or syntactic conditions. We have thus seen studies of uncountably categorical theories of groups [40, 43, 58, 7, 9, 69], rings [48, 59, 70, 71, 31, 6], theories of a one-place function [33], universal theories of semigroups [32], quasi-varieties categorical in infinite powers [1, 23, 24, 39, 50, 51, 66, 67], and Horn theories [25–27]. We shall refer to this research area as the special classification theory of categoricity.

The special classification theory has had considerable influence on our understanding of the nature of uncountable categoricity and the development of the general theory of categoricity, and has inspired several interesting conjectures. One result of this special trend has been an understanding of the fact that each of the known uncountably categorical structures is in a certain sense very close to one of the structures listed in the following three classes:

(1) algebraically closed fields, simple algebraic groups over algebraically closed fields, full rings of matrices over algebraically closed fields;

(2) vector spaces over countable and finite division rings, divisible abelian groups with finitely many elements of each finite order;

(3) the integers with the successor operation and, in greater generality, the graph of an arbitrary countable group.

At the same time, a general theory has emerged. Marsh [62], Baldwin and Lachlan [38], and, independently in a particular case, Palyutin [22] have observed that an important role in the study of uncountably categorical structures is played by their strongly minimal subsets, considered together with the algebraic closure operator. Baldwin in fact pointed out that when the entire structure is "rigidly" associated with a strongly minimal set, i.e., each element of the structure is in the algebraic closure of the strongly minimal set (almost strongly minimal structure), many important properties of the structure may be derived from those of the strongly minimal set [35, 36]. The way was thus paved for a separate theory of strongly minimal subsets of uncountably categorical structures and their relationship to the structure as a whole.

These notions and results have provided the foundations for the structural theory of categoricity, which has led to the solution of a number of important problems in categoricity theory; worthy of mention in this connection are [35, 36, 53, 38, 30, 5, 3, 4].

The aim of this dissertation is to develop a structural theory of categoricity, using methods and results of the special classification theory, and to construct, on this basis, foundations for a general classification theory of categoricity, that is to say, a theory aimed at describing large classes of uncountably categorical structures not restricted by any syntactic or algebraic conditions.

Underlying this dissertation is the idea that these two approaches may be combined in a productive manner.

We should mention that the connection between one particular problem of classification (the existence of a countably categorical pseudoplane) and certain general questions in model theory (stability theory) was first pointed out by Lachlan [52].

Going over to more rigorous formulations, we point out that in this introduction only the most necessary and less familiar concepts will be recalled. Other definitions and refinements may be found in the main text, in particular, with the aid of the index of definitions and notation.

The algebraic closure operation acl on a strongly minimal set A defines a dependence relation in the sense of van der Waerden's axioms [78], which generalize the properties of linear dependence in vector spaces and of algebraic dependence in fields. Such systems with closure are now known as (combinatorial) pregeometries or, if $\mathrm{acl}(a) = \{a\}$ for any point and $\mathrm{acl}(\varnothing) = \varnothing$, geometries [2]. Any pregeometry A can be converted into a geometry $\widehat{A}$ by removing the points $\mathrm{acl}(\varnothing)$ and identifying points that have the same closure. A projective geometry (over a division ring K) is a geometry isomorphic to the geometry $\widehat{V}$ obtained from the pregeometry of some vector space V (over K). Given a pregeometry A one can obtain a new pregeometry A_X for $X \subseteq A$ by defining a new closure operation acl_X on A as follows: $\mathrm{acl}_X(Y) = \mathrm{acl}(X \cup Y)$ for all $Y \subseteq A$. A pregeometry A is said to be locally projective if the geometry $\widehat{A_X}$ is a projective geometry for every singleton $X \subseteq A - \mathrm{acl}(\varnothing)$.

The pregeometries of the structures listed above in class (2) are locally projective. It is also not hard to see from Ziegler's results [79] about uncountably categorical modules that the pregeometries of strongly minimal sets in such modules are also locally projective. This motivates the definition: an uncountably categorical structure $\mathfrak{M}$ is said to be module-like if the pregeometries of any strongly minimal sets definable in $\mathfrak{M}$ are locally projective.

We shall say that an uncountably categorical structure $\mathfrak{M}$ is of disintegrated type if the geometry $\widehat{A}$ of any strongly minimal set A definable in $\mathfrak{M}$ is degenerate, i.e., $\mathrm{acl}(X \cup Y) = \mathrm{acl}(X) \cup \mathrm{acl}(Y)$ for any $X, Y \subseteq A$. All the examples in class (3) are structures of disintegrated of type.

To single out properties characteristic of the examples in class (1), we use the notion of a pseudoplane, which Lachlan [52] introduced in his study of certain questions pertaining to countably categorical structures. A pseudoplane is a structure whose underlying set is divided into two subsets—points and lines, with an incidence relation between elements of these two subsets, satisfying the following conditions: every point (line) is incident with an infinite set of lines (points); any two distinct points (lines) are incident with a finite set of common lines (points).

In all structures of class (1) one can define a pseudoplane more precisely as an affine plane over the corresponding algebraically closed field.

For this reason, structures in which a pseudoplane is definable will be called field-like structures. The main result of Chapter 2 is as follows:

THEOREM 2.3.1 (Trichotomy Theorem). *Any uncountably categorical structure is a structure of one of the three types: field-like, module-like, or of disintegrated type.*

This trichotomy in a sense also classifies structures according to "degrees of complexity". This is illustrated by the following results concerning groups, which will play an essential role in our study of the global structure of uncountably categorical structures.

THEOREM 2.4.1. *Let G be a group definable in an uncountable module-like categorical structure. Then,*

(a) *G is an almost abelian group*;

(b) *if G is infinite and has no infinite definable proper subgroups, then G is strongly minimal.*

THEOREM 2.4.5. *There exist no infinite groups definable in uncountably categorical structures of disintegrated type.*

Note that in the examples of class (1)—field-like structures—any algebraic groups over algebraically closed fields are definable. In other words, the complexity of the groups decreases from (1) to (3).

Chapter 3 of this dissertation is devoted to determining the position of totally categorical structures in our classification. The main result is as follows:

THEOREM 3.0.2. *There exist no totally categorical pseudoplanes. Totally categorical structures are module-like or of disintegrated type.*

This theorem will be proved via a profound combinatorial analysis of certain pregeometries and pseudoplanes. In the process, specific characteristics of definable subsets of totally categorical structures, found in Theorem 1.5.5, will be used in the traditional combinatorial calculations. More precisely: under certain natural restrictions, the latter theorem assigns to any subset of a totally categorical structure a polynomial with rational coefficients in one variable, which may be treated as the "number of elements" of the subset. It is these polynomials that will take part in our combinatorial calculations.

On the other hand, this assignment of polynomials to definable subsets yields a certain rank function with values in a polynomial ring. The classical Morley rank turns out to be a derivative of this rank; more precisely, the

Morley rank coincides with the degree of the polynomial. This sheds new light on the nature of the Morley rank.

Theorems 3.0.2 and 2.4.1 at once imply one of the most interesting results proved in recent years in model-theoretic algebra:

THEOREM (Baur-Cherlin-Macintyre [40]). *Totally categorical groups are almost abelian.*

Modulo some fairly simple arguments (Proposition 2.1.1 and Fact 2.1.2), Theorem 3.0.2 is equivalent to the following purely combinatorial theorem.

THEOREM 3.0.1. *A nondegenerate infinite locally finite homogeneous geometry is isomorphic to an affine or projective geometry of infinite dimension over a finite field.*

It is interesting to note that, independently of the realization of the importance of Theorems 3.0.1 and 3.0.2 for model theory [11], an equivalent conjecture was published in 1985 by specialists in permutation groups [86]. Their interest was motivated by the connection between these problems and the theory of Jordan groups and, in greater generality, the theory of doubly transitive groups. (The group of automorphisms of a homogeneous geometry is doubly transitive.)

Our proof of Theorems 3.0.1 and 3.0.2 was obtained in 1980 and published in 1981–1984 [13, 16, 17]. Simultaneously, the news spread that the classification of finite simple groups and, as a corollary, of doubly transitive groups, had finally been accomplished. Cherlin, being acquainted only with [11], pointed out that Theorems 3.0.1 and 3.0.2 follow from the classification of doubly transitive groups [44]. Independently, P. Neumann, also relying on the classification of finite simple groups, proved a weak version of Theorem 3.0.1 [86].

It is an important point that our proof (see above) makes no use of the classification of finite simple groups and is in fact self-contained. Moreover, this proof demonstrates the possibility of applying ideas of model theory in some questions of algebra.

In 1984 D. Evans, using ideas of the present writer, found a more direct combinatorial proof (using rather delicate considerations) of Theorem 3.0.1, and slightly later presented a finite-dimensional variant (for dimension at least 23) of the theorem [85].

The methods developed in Chapter 3 are also applicable in situations where the theory of finite groups and finite calculations are inapplicable. This is illustrated, in particular, by the results of Chapter 4, which we now proceed to formulate.

Marczewski [6] and Narkiewicz [65] introduced a class of closure algebras called v^*-algebras. Urbanik succeeded in describing all v^*-algebras of dimension greater than 2 [75]. Any such algebra must be one of the following:

(a) a vector space over a division ring with subspace of constants;

(b) an affine space over a division ring with a set of distinguished parallel translations;

(c) a set with a group of invertible unary operations on it, on condition that the sets of fixed points of any two operations other than the identity coincide.

Urbanik's theorem has been used repeatedly in model theory ([24–27, 51]), since certain strongly minimal subsets in models of uncountably categorical Horn theories turn out to be v^*-algebras. However, there are larger classes of theories whose strongly minimal subsets are not v^*-algebras, although they are very similar to them. We call a structure $\mathfrak{M}$ a *quasi-v^*-algebra* if $\mathrm{dcl}(X) = \mathrm{acl}(X)$ for any $X \subseteq M$, where $\mathrm{dcl}(X)$ is the definable closure of X, i.e., the union of all X-definable singletons.

Chapter 4 is devoted to a description of strongly minimal quasi-v^*-algebras. We note that every strongly minimal v^*-algebra is a quasi-v^*-algebra. The converse is not true, but an analog of Urbanik's theorem is possible.

We shall say that a structure $\mathfrak{M}_1$ is *rationally equivalent to a structure* $\mathfrak{M}_2$ *up to constants* if

$$M_1 - \mathrm{dcl}_{\mathfrak{M}_1}(\varnothing) = M = M_2 - \mathrm{dcl}_{\mathfrak{M}_2}(\varnothing)$$

and any relation on M obtained as the restriction of a $\varnothing$-definable relation in $\mathfrak{M}_1$ (resp. $\mathfrak{M}_2$) is the restriction of a $\varnothing$-definable relation in $\mathfrak{M}_2$ (resp. $\mathfrak{M}_1$).

THEOREM 4.3.10. *A strongly minimal quasi-v^*-algebra is rationally equivalent up to constants to one of the structures* (a)–(c) *of Urbanik's theorem.*

Theorem 4.3.10 considerably generalizes the strongly minimal version of Urbanik's theorem. Its proof is quite different from Urbanik's proof. Incidentally, our proof may also be modified for quasi-v^*-algebras of finite dimension, but this goes beyond the border lines of model theory and the topic of this dissertation.

There is an important application of Theorem 4.3.10 in the case of an uncountably categorical $\forall$-axiomatizable structure $\mathfrak{M}$. In that case it is easy to show that on strongly minimal subsets of $\mathfrak{M}$ one obtains the quasi-v^*-algebras described by Theorem 4.3.10. This also describes, to a significant degree, the structure of $\mathfrak{M}$ as a whole, since $\mathfrak{M}$ may be obtained as the closure of an arbitrary strongly minimal subset relative to all terms. A useful corollary of Theorem 4.3.10 is the following:

Any uncountably categorical $\forall$-axiomatizable structure is module-like or of disintegrated type.

This is not true, incidentally, for $\forall\exists$-axiomatizable structures.

Chapter 5 of this dissertation is devoted to the study of the global structure of uncountably categorical structures. Theorems 5.0.1–5.0.3, which we call "ladder theorems", describe this structure and are then used in the sequel.

The general case of the Ladder Theorem indicates the role of groups which are definable in uncountably categorical structures.

THEOREM 5.0.1. *Let* M_0 *and* M *be* $\varnothing$*-definable sets in an uncountably categorical structure, with* M_0 *strongly minimal. Then there exists a finite sequence of* $\varnothing$*-definable sets* $M_0, \dots, M_k$, *ending with the set* M, *such that for every* i $(0 \le i < k)$, *for every* M_i*-atom* $A \subseteq M_{i+1}$ *the group* $\mathrm{Gal}(A, M_i)$ *of all automorphisms of* A *which are elementary over* M_i *is definable together with its action on* A.

It is a useful observation that, since $\mathrm{Gal}(A, M_i)$ acts transitively on A, it follows that $R(A) \le R(\mathrm{Gal}(A, M_i))$. In particular, $M_{i+1} \subsetneq \mathrm{acl}(M_i)$ if and only if $\mathrm{Gal}(A, M_i)$ are finite for all M_i-atoms $A \subseteq M_{i+1}$.

Using Theorem 2.4.1, which is concerned with groups, one obtains the following version of the Ladder Theorem for module-like structures.

THEOREM 5.0.2. *Let* M_0, M *be* $\varnothing$*-definable sets in an uncountably categorical module-like structure,* $R(M_0) \ge 1$. *Then there exists a finite sequence of* $\varnothing$*-definable sets* $M_0, \dots, M_k$, *ending with the set* M, *such that* $R(A) \le 1$ *for every* M_i*-atom* $A \subseteq M_{i+1}$ *and* $i \in \{1, \dots, k-1\}$.

In other words, in module-like structures one can construct "ladders with rungs of height 1".

In the case of structures of disintegrated type, in view of Theorem 2.4.5, the ladder theorem degenerates to the following form:

THEOREM 5.0.3. *In an uncountably categorical structure* $\mathfrak{M}$ *of disintegrated type* $\mathrm{acl}(M_0)$ *contains all the elements of the structure, whatever the* $\varnothing$*-definable infinite set* M_0. *In particular,* $\mathfrak{M}$ *is almost strongly minimal.*

Using Theorems 5.0.2 and 5.0.3, we have established the following important property of totally categorical structures:

THEOREM 5.4.4 (finite submodel property). *If a sentence* α *is true in a totally categorical structure* $\mathfrak{M}$, *then* α *is also true in some finite substructure of* $\mathfrak{M}$.

An immediate corollary of this theorem is

THEOREM 5.4.5. *There exists no complete finitely axiomatizable totally categorical theory.*

This theorem completes the answer to a celebrated question: does there exist an uncountably categorical finitely axiomatizable theory? This question was first published in Vaught's 1963 survey and, as quickly became clear, could be divided into two different versions:

(a) Does there exist an uncountably categorical finitely axiomatizable theory which is not categorical in a countable power?

(b) Does there exist a complete finitely axiomatizable totally categorical theory?

The answer to (a) is positive, as was shown by Peretyat′kin [28], who constructed an example of such a theory. The answer to (b) is given by Theorem 5.4.5.

Several remarks on the history of the theorems cited above. The first step toward a proof of Theorem 5.4.4 was made by Makowsky [60], who proved it for almost strongly minimal theories. He observed that in that case any substructure of $\mathfrak{M}$ whose underlying set is $\mathrm{acl}(X)$ for a sufficiently large finite X is a finite model of α. In 1977 the present author [9] established the Ladder Theorem 5.0.1 (in an implicit form) and proved that if part (b) of Theorem 2.4.1 is true for totally categorical structures, then Theorem 5.4.4 is true. Somewhat later he proved that Theorem 3.0.2 implies one version of Theorem 3.0.1 and the latter implies 2.4.1(b) (announced in [8], full proof in [11]). Thus, by 1977 it had been proved that

$$3.0.2 \Rightarrow 3.0.1 \Rightarrow 2.4.1(\mathrm{b}) \Rightarrow 5.4.4.$$

The paper [11] also contained a proof of the main part of the Trichotomy Theorem for totally categorical theories.

In 1979 I pointed out the truth of Theorem 1.5.5 (Polynomial Theorem) and its significance for the solution of the problems discussed here. In [81] I published a proof of Theorem 5.4.4, using, as in previous papers, the Ladder Theorem 5.0.1, as well as the Polynomial Theorem. Here I also first introduced the notion of the envelope $\mathrm{en}(X)$ of a set X which, as it turned out, incorporates the properties of the algebraic closure $\mathrm{acl}(X)$ and of a prime submodel over X. Unfortunately, that proof involved an error, due to an inaccuracy in the formulation of the Polynomial Theorem, which left an essential gap in the proof. This gap was easily filled by using Theorem 3.0.1, which at the time was in fact already known to Cherlin and myself [13] (see above, comments on Theorems 3.0.1, 3.0.2). Cherlin, Harrington, and Lachlan did more: using Theorem 3.0.1 and envelopes, they proved Theorem 5.4.4 for countably categorical ω-stable theories and also proved that the Morley rank is finite in such theories. At the same time, the present writer showed that his original proof of Theorem 5.4.4 could be corrected without using such strong tools as Theorem 3.0.1. The correction [14] was based on a more systematic application of the Polynomial Theorem 1.5.5, using Siegel's Theorem on exponential Diophantine equations. This version of the proof of Theorem 5.4.4 is also presented below (§5, Chapter 5). We note that Theorem 1.5.5 also plays a key role in our proof of Theorem 3.0.2.

Conclusions. The results presented herein constitute a significant development of structural categoricity theory (Theorems 2.3.1, 5.0.1, 5.0.2, 5.0.3, 5.4.4), as well as a number of exhaustive classification theorems (3.0.1, 3.0.2, 4.3.10), which yield as particular cases such special classification results as the theorems of Baur-Cherlin-Macintyre [40] and Urbanik (the strongly minimal case) [75]. Theorem 3.0.1 answers a question that arose in the theory of permutation groups [86]. The theory constructed permitted the solution of the well-known problem on the impossibility of finite axiomatizability of a complete totally categorical theory (Theorem 5.4.5).

Application and further development. The ideas of Theorem 2.3.1 are developed and applied in [42, 68]. Theorem 3.0.1 is used in [34] to generalize Theorem 5.0.3. Theorem 3.0.1 and the envelopes of Chapter 5, §4 are used in [44] to study ω-stable ω-categorical theories and to generalize Theorems 5.4.5 and 3.0.2. A further investigation of envelopes may be found in [57]. Theorem 2.4.1 is generalized in [84]. Theorem 3.0.1 is strengthened in [85].

Organization of the text. This dissertation is divided into chapters, the chapters into sections. A number $n.m.k$ applied to an assertion (theorem, etc.) means the kth assertion (theorem, etc.) in § m of Chapter n. The enumeration of formulas is similar (but in parentheses).

CHAPTER I

Survey of Preliminary Results and Terminology

The goal of this chapter, in addition to explaining the notation and terminology, is to survey the results of categoricity theory on the basis of which the subsequent theory is constructed. The major part of these results is known and may be found in the quite accessible sources [21, 29]. Such results will be cited without indicating any source and are called "facts". "Facts" of other kinds, less well known, will be cited together with their original sources. Finally, the chapter also includes proved results, lemmas, propositions, and a theorem. Some of them are very simple and essentially known. Instead of the proof in such cases we provide only hints. Theorem 1.5.5, which is due to the author, is an important tool for the study of totally categorical structures and may be counted as one of the main results of this dissertation.

§1. Notation and prerequisites about uncountably categorical theories

By the word *structure* indexStructure we will always mean a structure of a countable language without function symbols (unless the opposite is evident). Structures will be denoted by capital Gothic letters, their underlying sets by the parallel Latin letters. The signature of a structure $\mathfrak{M}$ will be denoted by $\Omega_{\mathfrak{M}}$ (possibly without subscript). If $X \subseteq M$, then $\Omega_{\mathfrak{M}}(X)$ is the enrichment ([1]) of $\Omega_{\mathfrak{M}}$ by constant symbols for the elements of X. The structure $\mathfrak{M}$ will almost always be fixed, and $F_n(X)$ will denote the set of formulas of signature $\Omega_{\mathfrak{M}}(X)$ with n free variables, $F_n = F_n(\varnothing)$, $F(X) = \bigcup_n F_n(X)$, $F = F(\varnothing)$. Taking licence, we shall consider $F_n(X)$ as a Boolean algebra with operations $\vee$, $\&$, $\neg$. This Boolean algebra is isomorphic to the Boolean algebra of subsets of M^n which are definable by formulas relative to X (X-*definable subsets*). $S_n(X)$ is the set of complete n-types over X, i.e., the Stone space of the Boolean algebra $F_n(X)$. If $\bar{x} = \langle x_1, \dots, x_n \rangle$, then $|\bar{x}|$ denotes the set $\{x_1, \dots, x_n\}$ and instead of "$|\bar{x}|$-definable" we shall say "$\bar{x}$-definable". The atoms of the algebra $F_n(X)$ will be called X-*atoms*. The $\varnothing$-definable atoms will be called simply atoms, and "*definable subset of* M^n" will be short for "M-definable subset of M^n".

For formulas $\varphi(v_1, \dots, v_k, w_1, \dots, w_n)$ in F_{k+n} and elements $x_1, \dots,$

([1]) or expansion. Editor's note.

x_k we denote

$$\varphi(x_1, \dots, x_k, \mathfrak{M}) = \{\langle y_1, \dots, y_n\rangle \in M^n : \mathfrak{M} \models \varphi(x_1, \dots, x_k, y_1, \dots, y_n)\}.$$

Another system of notation for definable subsets will be used when we are working with binary definable relations. Let I be such a relation; xIy will stand for the fact that $\langle x, y\rangle \in I$, and

$$xI = \{y \in M : xIy\}, \qquad Iy = \{x \in M : xIy\}.$$

In particular, if E is an equivalence relation, then $xE = Ex$ is the equivalence class with representative x.

As a rule, i, j, k, l, m, n will denote natural numbers, λ, μ, κ infinite cardinals, ω the set of natural numbers; $\|X\|$ will denote the cardinality of a set X.

The following remark frequently proves useful in establishing the definability of subsets relative to different sets of constants.

PROPOSITION 1.1. *Let $\mathfrak{M}$ be a λ-saturated λ-homogeneous structure, $X, Y \subseteq M$, $\varphi \in F_k(X)$, $\|Y\| < \lambda$, and let $\varphi(\mathfrak{M})$ be stable (invariant) under all automorphisms of M that are the identity on Y. Then there exists a formula $\varphi' \in F_k(Y)$ such that $\varphi'(\mathfrak{M}) = \varphi(\mathfrak{M})$.*

The proof amounts to a standard application of the Compactness Theorem.

FACT 1.2. *Any uncountable model of an uncountably categorical theory is saturated (and consequently also homogeneous) in its own cardinality.*

One says that *a structure $\mathfrak{M}$ is atomic over X* if every tuple $\bar{y}$ of elements of $\mathfrak{M}$ satisfies some X-atom; *$\mathfrak{M}$ is prime over X* if any elementary monomorphism of X into any structure $\mathfrak{M}'$ extends to an elementary embedding of $\mathfrak{M}$ into $\mathfrak{M}'$.

FACT 1.3. *If $\mathfrak{M}$ is an uncountably categorical structure and A an infinite definable subset of M, then $\mathfrak{M}$ is prime and atomic over A.*

A set X is said to be *indiscernible over X_0* if any bijection of X onto itself is an X_0-elementary monomorphism.

FACT 1.4. *If $\mathfrak{M}$ is an uncountably categorical structure of cardinality $\lambda \geq \aleph_0$, A an infinite X_0-definable subset of M, and $\|X_0\| < \lambda$, then A contains a subset of cardinality λ indiscernible over X_0.*

FACT 1.5. *For any model $\mathfrak{M}$ of an uncountably categorical theory T*

$$\|S_n(M)\| = \|M\|;$$

in other words, T is ω-stable.

An infinite definable subset $\varphi(\mathfrak{M})$ of a structure $\mathfrak{M}$ is said to be *strongly minimal* if $\varphi(\mathfrak{M}')$ cannot be split into two infinite definable subsets for any elementary extension $\mathfrak{M}'$ of $\mathfrak{M}$.

FACT 1.6. *A theory T is uncountably categorical if and only if T is ω-stable and for any model $\mathfrak{M}$ of T and any strongly minimal subset A of $\mathfrak{M}$ the cardinalities of A and M are equal.*

This easily implies

FACT 1.7. *If a structure $\mathfrak{M}$ is uncountably categorical, then any enrichment of $\mathfrak{M}$ by a countable set of constants is also uncountably categorical.*

§2. Rank and independence

Throughout, "rank" means "Morley rank". $R(p)$ will denote the rank of a type p in the Stone space $S_n(X)$, $R(\varphi)$ the rank of a formula (a definable set) $\varphi(\mathfrak{M})$, which is defined as the maximum of $R(p)$ over those p such that $\varphi \in p$.

Morley's original definition of $R(p)$ [64] may be replaced by the following topological variant, if the structure under consideration is sufficiently saturated.

FACT 2.1. *If $\mathfrak{M}$ is countably saturated, $p \in S_n(M)$ and α is an ordinal, then $R(p) > \alpha$ if and only if p is a limit point for the set of types of rank α in $S_n(M)$ in the Stone topology. $R(\varphi) = 0$ if φ is valid only for a finite set of elements in M.*

By $\mathrm{RD}(\varphi)$ we denote the pair $\langle r, d\rangle$, where $r = R(\varphi)$ and $d = D(\varphi)$ is the degree of φ, i.e., the largest natural number for which there exists a partition of φ into d definable subsets of rank r. As follows from Fact 2.1, RD is an invariant of the Boolean algebra $F_n(M)$, and in fact more—an invariant of the Boolean algebra of definable subsets of the set $\varphi(\mathfrak{M})$, if $\mathfrak{M}$ is countably saturated.

Throughout the sequel we shall assume that $\mathfrak{M}$ is a "big", sufficiently saturated structure.

FACT 2.2 ([37, 5]). *For any definable set A in an $\aleph_1$-categorical structure, $R(A)$ is finite.*

A formula $\psi(v_0, v_1)$ is said to be a *stratification* of the set defined by a formula $\varphi(v_1)$ with respect to the set defined by a formula $\delta(v_0)$ if

$$\mathfrak{M} \models \psi(v_0, v_1) \to \delta(v_0) \quad \text{and} \quad \mathfrak{M} \models (\exists v_0)\psi(v_0, v_1) \leftrightarrow \varphi(v_1).$$

The subsets $\psi(x, \mathfrak{M})$, where $x \in \delta(\mathfrak{M})$, are called *strata*. If the rank of each stratum is at most m, then ψ is a *stratification of rank at most m*. If the rank of each stratum is exactly m and

$$\mathfrak{M} \models \psi(v_0, v_1) \,\&\, \psi(v_0', v_1) \to v_0 = v_0',$$

then ψ is called an *exact stratification of rank m*.

FACT 2.3 ([5]). *For any infinite sets* A *and* B *definable in an* $\aleph_1$*-categorical structure, there exists a stratification of* A *with respect to* B *of rank at most* $R(A)-1$, *which is definable over* $A \cup B$.

It is easy to see that this assertion is equivalent to the following

FACT 2.4 ([73, 74]). *For any infinite set* B *definable in an* $\aleph_1$*-categorical structure* $\mathfrak{M}$, *there is no infinite set in* $\mathfrak{M}$ *indiscernible over* B.

FACT 2.5 ([5, 74]). *If* ψ *is a stratification of rank at most* m *of a set* A *with respect to a set* B *in an* $\aleph_1$*-categorical structure, then*

$$R(A) \leq m + R(B).$$

If ψ *is an exact stratification of rank* m, *the above inequality becomes an equality.*

FACT 2.6 ([5, 37, 74]). *For any formula* $\psi \in F_2$ *in an* $\aleph_1$*-categorical structure* $\mathfrak{M}$ *and any natural number* m, *there exists a formula* $\vartheta_m \in F_1$ *such that*

$$R(\psi(x, \mathfrak{M})) \leq m \Leftrightarrow x \in \vartheta_m(\mathfrak{M}).$$

Given definable sets A and B, we write $A \sqsupset B$ (read: *A almost contains B*) to denote the fact that $R(B-A) < R(B)$. $A \boxminus B$ denotes $A \sqsupset B \,\&\, B \sqsupset A$.

It follows from Facts 2.2 and 2.6 that for any two formulas $\psi_1, \psi_2 \in F_2$ in an $\aleph_1$-categorical structure $\mathfrak{M}$ there exists a formula $\vartheta \in F_2$ such that

$$\psi_1(x, \mathfrak{M}) \sqsupset \psi_2(y, \mathfrak{M}) \Leftrightarrow \mathfrak{M} \models \vartheta(x, y).$$

PROPOSITION 2.7. *Suppose that in an* $\aleph_1$*-categorical structure* $\mathfrak{M}$ *there exist a formula* $\gamma(v_0, v_1)$ *and definable sets* A *and* B *such that for any finite subset* X *of* A *there is an element* $x \in B$ *such that*

$$\gamma(x, \mathfrak{M}) \supseteq X.$$

Then there exists $y \in B$ *such that*

$$\gamma(y, \mathfrak{M}) \sqsupset A.$$

PROOF BY INDUCTION ON $R(A)$. If $R(A) = 0$, then A is a finite set and the assertion is true by assumption.

Let A be strongly minimal. By assumption and by the Compactness Theorem, there exists $y \in B$ such that $\gamma(y, \mathfrak{M}) \cap A$ is infinite. Since A is strongly minimal,

$$\gamma(y, \mathfrak{M}) \sqsupset A.$$

Now let $R(A) = m+1$ and suppose that the assertion is true for definable sets of rank at most m. Consider a stratification $\psi(v_0, v_1)$ of rank at most m of A with respect to an arbitrary strongly minimal set C. By the inductive hypothesis, for any finite $Z \subseteq C$ there exists an element $z \in B$ such that

$$\gamma(z, \mathfrak{M}) \sqsupset \bigcup_{c \in Z} \psi(c, \mathfrak{M}).$$

Let $\vartheta(v_0, v_1)$ denote a formula for which

$$R(\psi(x, \mathfrak{M}) - \gamma(y, \mathfrak{M})) \leq m - 1 \Leftrightarrow \mathfrak{M} \models \vartheta(y, x).$$

By definition, for any finite $Z \subseteq C$

$$\vartheta(z, \mathfrak{M}) \supseteq Z;$$

therefore, as shown above, for strongly minimal C there exists $y \in B$ such that

$$\vartheta(y, \mathfrak{M}) \sqsupset C,$$

that is, $C - \vartheta(y, \mathfrak{M})$ is finite. Now, evaluating the rank with the help of Fact 2.5, we obtain

$$R(A - \gamma(y, \mathfrak{M})) \leq m. \quad \square$$

A convenient tool for working with ranks was proposed by Lascar [54]. Let x be an element and X a subset of a structure $\mathfrak{M}$. Then $\mathrm{tp}(x, X)$ will denote the complete type over X realized by x. Instead of $R(\mathrm{tp}(x, X))$ we write $R(x, X)$.

The Morley rank in an $\aleph_1$-categorical structure $\mathfrak{M}$ has the following connectivity property: *for any type $p \in S_n(X)$ there exist $X' \supseteq X$, such that X' is a subset of some elementary extension of $\mathfrak{M}$, and a type $p' \in S_n(X')$, $p' \supseteq p$, $R(p') = R(p) - 1$, if $R(p) > 0$.*

Let us prove this. Let $\varphi \in p$, $R(\varphi) = R(p)$, and let $\delta(v)$ be a strongly minimal formula. Consider a stratification ψ of rank at most $R(p) - 1$ of the set $\varphi(\mathfrak{M})$ with respect to $\delta(\mathfrak{M})$. We may assume that ψ and δ are X-definable. It follows from Fact 2.5 that for all elements $x \in \delta(\mathfrak{M})$ $R(\psi(x, \mathfrak{M})) = R(p) - 1$, except for possibly finitely many of them. Now select in some elementary extension $\mathfrak{M}'$ an element $x_0 \in \delta(\mathfrak{M}') - \delta(\mathfrak{M})$; then $R(\psi(x_0, \mathfrak{M}')) = R(p) - 1$. Since for any formula $\varphi' \in p$ such that $\varphi'(\mathfrak{M}) \subseteq \varphi(\mathfrak{M})$, the formula $\psi(v_0, v_1) \,\&\, \varphi'(v_1)$ is a stratification of rank at most $R(p) - 1$ of the set $\varphi'(\mathfrak{M})$ with respect to $\delta(\mathfrak{M})$, it follows as before that $R(\psi(x_0, \mathfrak{M}') \cap \varphi'(\mathfrak{M}')) = R(p) - 1$. This means that $R(p \cup \{\psi(x_0, v_1)\}) = R(p) - 1$. Setting $X' = X \cup \{x_0\}$ and letting p' be a complete type of maximal rank over X' extending $p \cup \{\psi(x_0, v_1)\}$, we obtain the desired result. $\square$

Thus, the Morley rank in uncountably categorical structures satisfies the following relation, which is truc for any connected rank:

(1) $R(\langle x, y\rangle, Z) = R(x, \{y\} \cup Z) + R(y, Z) = R(y, \{x\} \cup Z) + R(x, Z)$,

which implies the following Reciprocity Principle:

(2) $R(x, \{y\} \cup Z) < R(x, Z) \Leftrightarrow R(y, \{x\} \cup Z) < R(y, Z)$.

Let A be a definable subset of M, $X \subseteq M$. We denote

$$A^{(n)}[X] = \{a \in A : R(a, X) \leq n\}, \qquad A[X] = A^{(0)}[X].$$

The set $M[X]$ will play an important role; we denote it by $\mathrm{acl}(X)$ and call it the *algebraic closure* of X. Thus,

$$A[X] = \mathrm{acl}(X) \cap A.$$

It follows from (2) that

(3) $R(x, Z) = R(x, \operatorname{acl}(Z))$.

It is easy to see that

(4) $\operatorname{acl}(X) = \bigcup\{\operatorname{acl}(X') : X' \subseteq X,\ X' \text{ finite}\}$;

(5) $\operatorname{acl}(X) \supseteq X$;

(6) $\operatorname{acl}(X) \supseteq Y \Rightarrow \operatorname{acl}(X) \supseteq \operatorname{acl}(Y)$.

If A is a strongly minimal set, then for any $y, z \in A$

(7) $y \in \operatorname{acl}(\{z\} \cup X) - \operatorname{acl}(X) \Rightarrow z \in \operatorname{acl}(\{y\} \cup X)$;

(8) $y, z \notin \operatorname{acl}(X) \Rightarrow$ there exists an elementary bijection $f: A \to A$ such that $f(y) = z$, $f|_X = \operatorname{id}_X$.

The properties listed above enable us to consider algebraic independence of elements of A, *basis* and *dimension* of closed subsets X (the latter is denoted by $\dim X$) of a strongly minimal set A.

Let A, B be definable subsets of $\mathfrak{M}$, $X \subseteq \mathfrak{M}$. We say that B is *coordinatizable over* A_X if

$$B \subseteq \operatorname{acl}(A \cup X).$$

If $x = \varnothing$, we shall refer simply to coordinatization over A.

We shall say that B is *k-coordinatizable* over A_X if for any $b \in B$ there exist $a_1, \dots, a_k \in A$ such that

$$b \in \operatorname{acl}(\{a_1, \dots, a_k\} \cup X).$$

In the case that for any $b \in B$ one can select $a_1, \dots, a_k \in A$ so that

$$\operatorname{acl}(\{b\} \cup X) = \operatorname{acl}(\{a_1, \dots, a_k\} \cup X),$$

then B is *strictly k-coordinatizable* over A_X.

If $X = |\bar{x}|$, we speak of coordinatization over $A_{\bar{x}}$.

LEMMA 2.8. *If $\mathfrak{M}$ is a countably saturated structure and B is coordinatizable over A_X in $\mathfrak{M}$, where X is finite, then B is k-coordinatizable over A_X for some k.*

A structure $\mathfrak{M}$ is said to be *almost strongly minimal* if M is k-coordinatizable over A_X for some k, strongly minimal A and finite X.

It is readily proved that any almost strongly minimal structure has an uncountably categorical theory.

LEMMA 2.9. *If $\mathfrak{M}$ is an uncountably categorical structure, A is strongly minimal, and B is k-coordinatizable over A_X, where X is finite, then $R(B) \leq k$ and there exists a finite set $X' \supseteq X$ such that B is strictly $R(B)$-coordinatizable over $A_{X'}$.*

PROOF. We may assume that A and B are X-definable. Let $b \in B$, $R(b, X) = R(B) > 0$, $a_1, \dots, a_k \in A$, $b \in \operatorname{acl}(\{a_1, \dots, a_k\} \cup X)$, let $a_1, \dots, a_k$ be algebraically independent over X, $0 \leq j < k$, and $X_j = X \cup \{a_{j+1}, \dots, a_k\}$. Then, by (1),

$$R(b, X_j) = R(b, X_{j+1}) - (R(a_{j+1}, X_{j+1}) - R(a_{j+1}, \{b\} \cup X_{j+1})).$$

Since the expression in parentheses on the right is at most 1, we have

$$R(b, X_j) \geq R(b, X_{j+1}) - 1.$$

Hence $R(b, X_0) \geq R(b, X_k) - k$, but $X_k = X$, $R(b, X) = R(B)$, $R(b, X_0) = 0$, and therefore $R(B) \leq k$.

Now let j be the least number such that

$$R(b, X_j) = R(B)$$

for any arrangement of $a_1, \ldots, a_k$. Then

$$j > 0, \qquad R(b, \{a_i\} \cup X_j) < R(b, X_j),$$

and therefore

$$a_i \in \operatorname{acl}(\{b\} \cup X_j) \quad \text{for } i = 1, \ldots, j.$$

This at once implies that

$$\operatorname{acl}(\{a_1, \ldots, a_j\} \cup X_j) = \operatorname{acl}(\{b\} \cup X_j),$$

and by (1)

$$R(b, X_{j+i}) = R(b, X_j) - i,$$

that is to say,

$$0 = R(b, X_k) = R(B) - j, \qquad j = R(B).$$

There clearly exists an X_j-definable set B' containing b, each element y of which satisfies, following b, the condition

$$(\exists x_1, \ldots, x_j \in A)(\operatorname{acl}(\{x_1, \ldots, x_j\} \cup X_j) = \operatorname{acl}(\{y\} \cup X_j)),$$

that is, B' is strictly coordinatizable over A_{X_j}, $j = R(B)$. Since $b \in B'$, it follows that $R(B') = R(B)$, and so $(B - B')$ is strictly r-coordinatizable over A_Y for some $Y \supseteq X$, $r = R(B - B')$. Defining $X' = X_j \cup Y$, we complete the proof. □

Remark. This assertion is true without the assumption that $\mathfrak{M}$ is uncountably categorical.

Lascar [54] has introduced an independence concept for sets: X and Y are said to be *independent over* Z (X is independent of Y over Z) if

$$R(\bar{x}, Y \cup Z) = R(\bar{x}, Z)$$

for any tuple $\bar{x}$ of elements of X. Property (2) above implies that this relation is symmetric.

Lemma 2.10. *Let p be an n-type over Z and assume that X is independent of each of the sets $Y_1, \ldots, Y_n$ over Z. Then there exists a tuple $\bar{a} = \langle a_1, \ldots, a_n \rangle$ realizing p such that X is independent of each of the sets $|\bar{a}| \cup Y_1, \ldots, |\bar{a}| \cup Y_k$ over Z.*

Proof. Consider a complete extension p' of p over $X \cup Y_1 \cup \cdots Y_k \cup Z$ such that $R(p') = R(p)$. Let $\langle a_1, \ldots, a_n \rangle$ realize p'. Then

$$R(\bar{a}, X \cup Y_i \cup Z) = R(\bar{a}, Y_i \cup Z) \quad \text{for } i = 1, \ldots, k.$$

Hence, by the Reciprocity Principle,

$$R(\bar{x}, |\bar{a}| \cup Y_i \cup Z) = R(\bar{x}, Y_i \cup Z) = R(\bar{x}, Z)$$

for any tuple $\bar{x}$ of elements of X. □

§3. Definability

The concept of definability of subsets considered in the previous sections turns out to be overly narrow in some questions of structural categoricity theory. The broader definition given below was proposed in [74] and [9] and, in a less explicit form, in many earlier publications in model theory.

Let $\mathfrak{M}$ and $\mathfrak{N}$ be structures, $X \subseteq M$. We say that $\mathfrak{N}$ is *X-definable in* $\mathfrak{N}$ if there exist k and a map $\sigma: M^k \to N$ such that

(1) $\operatorname{Dom}\sigma$ is an X-definable subset of M^k;

(2) the preimages of equality and all predicates in $\Omega_{\mathfrak{N}}$ are X-definable in $\mathfrak{M}$.

Under these conditions we refer to the map σ as an *X-interpretation of* $\mathfrak{N}$ *in* $\mathfrak{M}$.

If $\Omega_{\mathfrak{N}} = \varnothing$, we shall say that the *set N is X-definable in* $\mathfrak{M}$. We may always assume, up to a bijection, that N is a quotient set A/E, where A is an X-definable subset and E an X-definable equivalence relation in $\mathfrak{M}$.

We shall omit X in the term "X-definable" if $X = M$.

LEMMA 3.1. *Let σ be an X-interpretation of a structure $\mathfrak{N}$ in a structure $\mathfrak{M}$. Then*

(a) *the σ-preimages of the $\varnothing$-definable predicates of $\mathfrak{N}$ are X-definable in $\mathfrak{M}$;*

(b) *if $\mathfrak{L}$ is a $\varnothing$-definable structure in $\mathfrak{N}$, then $\mathfrak{L}$ is X-definable in $\mathfrak{M}$;*

(c) *if $\mathfrak{M}$ is λ-saturated and $\lambda > \|X\|$, then $\mathfrak{N}$ is also λ-saturated.*

PROOF. Part (a) is a standard application of induction on complexity of formulas. Parts (b) and (c) are direct consequences of (a). □

Let σ be an X-interpretation of $\mathfrak{N}$ in $\mathfrak{M}$, $\Omega_{\mathfrak{M}} \cap \Omega_{\mathfrak{N}} = \varnothing$, $M \cap N = \varnothing$. Then an *extension of $\mathfrak{M}$ by definition of $\mathfrak{N}$ through σ* is a structure $\mathfrak{M}^\sigma\mathfrak{N}$ defined by the following conditions:

The signature of the structure $\mathfrak{M}^\sigma\mathfrak{N}$ is $\Omega_{\mathfrak{M}} \cup \Omega_{\mathfrak{N}} \cup \{P_M, P_N, P_\sigma\}$; the underlying set of $\mathfrak{M}^\sigma\mathfrak{N}$ is $M \cup N$;

$$P_M(\mathfrak{M}^\sigma\mathfrak{N}) = M, \quad P_N(\mathfrak{M}^\sigma\mathfrak{N}) = N, \quad P_\sigma(\mathfrak{M}^\sigma\mathfrak{N}) = \operatorname{graph}\sigma;$$

$$P(\mathfrak{M}^\sigma\mathfrak{N}) = P(\mathfrak{M}) \quad \text{for } P \in \Omega_{\mathfrak{M}};$$

$$P(\mathfrak{M}^\sigma\mathfrak{N}) = P(\mathfrak{N}) \quad \text{for } P \in \Omega_{\mathfrak{N}}.$$

Shelah [74] applies the term *extension of $\mathfrak{M}$ by imaginary elements* to a similar construction.

It follows directly from the definition that any structure which is elementarily equivalent to $\mathfrak{M}^\sigma\mathfrak{N}$ is also an extension of some structure $\mathfrak{M}'$ by definition of a structure $\mathfrak{N}'$ through σ', where $\mathfrak{M}' \equiv \mathfrak{M}$, $\mathfrak{N}' \equiv \mathfrak{N}$.

This implies

LEMMA 3.2. *If $\mathfrak{M}$ is an uncountably categorical structure, X is countable and σ is an X-interpretation of $\mathfrak{N}$ in $\mathfrak{M}$, then $\mathfrak{M}^\sigma\mathfrak{N}$ is uncountably categorical.*

The following lemma is also useful.

LEMMA 3.3. *$\mathfrak{M}^\sigma\mathfrak{N}$ is X-definable in $\mathfrak{M}$ if σ is an X-interpretation of $\mathfrak{N}$ in $\mathfrak{M}$ and M contains two distinct elements.*

PROOF. Define a map ρ with domain

$$\operatorname{Dom}\rho = M^3 \times \operatorname{Dom}\sigma$$

as follows:

$$\rho(x_1, x_2, y, \bar{z}) = \begin{cases} y & \text{if } x_1 = x_2, \\ \sigma(\bar{z}) & \text{if } x_1 \neq x_2. \end{cases}$$

ρ is an X-interpretation of $\mathfrak{M}^\sigma\mathfrak{N}$ in $\mathfrak{M}$. □

Note that for fixed $x \in M$, $\bar{z} \in \operatorname{Dom}\sigma$, the map $y \to \rho(x, x, y, \bar{z})$ is an embedding of M into $\mathfrak{M}^\sigma\mathfrak{N}$, induced by an isomorphism (by Lemma 3.1) between the Boolean algebras of M-definable subsets of M^k in $\mathfrak{M}$ and $\mathfrak{M}^\sigma\mathfrak{N}$.

COROLLARY. *For an M-definable subset $A \subseteq M^k$, the values of $R(A)$ in the sense of $\mathfrak{M}$ and $\mathfrak{M}^\sigma\mathfrak{N}$ coincide.*

LEMMA 3.4. *Let A be a $\varnothing$-definable subset of M^k in $\mathfrak{M}$ and assume that for each $\bar{x} \in A$ there exists an $\bar{x}$-interpretation $\sigma_{\bar{x}}$ of some set $N_{\bar{x}}$ in $\mathfrak{M}$, such that $N_{\bar{x}} \cap N_{\bar{x}'} = \varnothing$ for $\bar{x} \neq \bar{x}'$. Then there exists a $\varnothing$-interpretation σ of the set $N = \bigcup\{N_{\bar{x}} : \bar{x} \in A\}$ and $\sigma \supseteq \sigma_{\bar{x}}$ for all $\bar{x} \in A$.*

PROOF. Define

$$\operatorname{Dom}\sigma = \{\langle \bar{x}, y\rangle : \bar{x} \in A,\ y \in \operatorname{Dom}\sigma_{\bar{x}}\}$$

and

$$\sigma(\bar{x}, y) = \sigma_{\bar{x}}(y) \quad \text{for } \langle \bar{x}, y\rangle \in \operatorname{Dom}\sigma.$$

It is easily verified that σ is a $\varnothing$-interpretation of N in $\mathfrak{M}$. □

COROLLARY 3.5. *Every set N_X X-definable in $\mathfrak{M}$ is a subset, X-definable in $\mathfrak{M}$, of some $\varnothing$-definable set N in $\mathfrak{M}$. In addition, an X-interpretation σ_X for N_X extends to a $\varnothing$-interpretation σ for N.*

Indeed, let $X = \{x_1, \dots, x_n\}$, $\bar{x} = \langle x_1, \dots, x_n\rangle \in A \subseteq M^n$, and let A be $\varnothing$-definable,

$$N_X = N_{\bar{x}} = D_{\bar{x}}/E_{\bar{x}},$$

where $E_{\bar{x}}$ is an $\bar{x}$-definable equivalence on $D_{\bar{x}} = \operatorname{Dom}\sigma_X$,

$$D_{\bar{x}} = \varphi(\bar{x}, \mathfrak{M}), \quad E_{\bar{x}} = \psi(\bar{x}, \mathfrak{M}), \qquad \varphi, \psi \in F.$$

We may assume, diminishing A if needed, that

$$A \subseteq \{\bar{y} \in M^n : \psi(\bar{y}, \mathfrak{M}) \text{ is an equivalence on } \varphi(\bar{y}, \mathfrak{M})\}.$$

We now define $N_{\bar{y}} = \varphi(\bar{y}, \mathfrak{M})/\psi(\bar{y}, \mathfrak{M})$ for $\bar{y} \in A$, assuming that these sets are disjoint for different $\bar{y}$. Lemma 3.4 now yields the desired N. □

In view of the above assertions, we may consider X-definable sets in $\mathfrak{M}$ as X-definable subsets of some extension of $\mathfrak{M}$ by a definition. The extended structure inherits from $\mathfrak{M}$ its uncountable categoricity, definable subsets, rank, saturation. Throughout the sequel we shall make use of this convention; in particular, we shall consider the rank and definable subsets of a structure $\mathfrak{N}$, definable in $\mathfrak{M}$, in the sense of the structure $\mathfrak{M}^\sigma\mathfrak{N}$.

It should be noted that according to this convention tuples in M^n are elements of some extension of $\mathfrak{M}$ by a definition. In particular, this makes it meaningful to write, for $\bar{x} \in M^n$:

$$\bar{x} \in \operatorname{acl}(Y);$$

it is easy to see that this is equivalent to

$$|\bar{x}| \subseteq \operatorname{acl}(Y).$$

Moreover, we shall agree to consider formulas in extensions $\mathfrak{M}^\sigma\mathfrak{N}$ of a structure $\mathfrak{M}$ as formulas in $\mathfrak{M}$, thus writing $\mathfrak{M} \models \varphi$ instead of $\mathfrak{M}^\sigma\mathfrak{N} \models \varphi$.

One effective application of the definability concept considered above is demonstrated by the formulation of Shelah's Finite Equivalence Relation Theorem, presented below in a somewhat weakened version.

FACT 3.6. *On every X-definable set A in an ω-stable structure there exists an X-definable equivalence relation E such that A/E is finite and every equivalence class is either of rank less than that of A or of degree one. In particular, there exists an* $\operatorname{acl}(X)$*-definable subset $B \subseteq A$ such that* $\operatorname{RD}(B) = \langle R(A), 1\rangle$. *More precisely, B is one of the equivalence classes, B is b-definable for $b \in A/E \subseteq \operatorname{acl}(X)$.*

§4. Groups definable in uncountably categorical structures

Typical examples of groups which are definable in uncountably categorical structures are algebraic groups over algebraically closed fields. Among the groups which are not almost abelian this is the only known class. The theory evolved up to the present points to a profound analogy between the theory of algebraic groups over an algebraically closed field and the theory of groups definable in uncountably categorical structures. A survey follows of the facts about such groups necessary for the sequel.

Throughout, G will be a group definable in an uncountably categorical structure $\mathfrak{M}$. Recall that definability in G is understood in the sense of definability in the structure obtained by extending $\mathfrak{M}$ by definition of the group G.

FACT 4.1 ([59]). *An infinite field whose theory is ω-stable is algebraically closed.*

FACT 4.2 ([7, 43, 69]). *An infinite group G contains an infinite abelian definable subgroup. If G contains no proper infinite definable subgroups then G is an abelian group; in particular, strongly minimal groups are abelian.*

A group G is said to be *connected* if G contains no definable subgroups of finite index.

FACT 4.3 ([7, 43]). *G contains a unique connected definable subgroup G^0 of finite index in G; G^0 is $\varnothing$-definable if G is $\varnothing$-definable;* $R(G) = R(G^0)$.

The subgroup G^0 defined in Fact 4.3 is called the connected component of G.

FACT 4.4 ([7, 43]). *A finite normal subgroup of a connected group G lies in the center of G.*

Let $A \subseteq G$, $g \in G$. We denote

$$g \cdot A = \{g \cdot a : a \in A\}, \qquad A \cdot g = \{a \cdot g : a \in A\}.$$

The maps $a \to g \cdot a$, $a \to a \cdot g$, with g fixed, define automorphisms of the Boolean algebra $F_1^G(M)$ of definable subsets of G. These automorphisms induce a homeomorphism of $S_1^G(M)$, the Stone space of $F_1^G(M)$, onto itself. For $p \in S_1^G(M)$, $g \in G$, we denote

$$g \cdot p = \{g \cdot A : A \in p\}, \qquad p \cdot g = \{A \cdot g : A \in p\}.$$

As already pointed out, $g \cdot p$ and $p \cdot g$ are complete types.

Let

$$\mathrm{St}_G(p) = \{g \in G : g \cdot p = p\}$$

denote the (left) stabilizer of type p in G.

FACT 4.5 ([7]). *The stabilizer of complete type p is a definable subgroup of G. Moreover, if $A \in p$,* $\mathrm{RD}(A) = \mathrm{RD}(p)$, *then*

$$\mathrm{St}_G(p) = \{g \in G : g \cdot A \sqcap A\}.$$

FACT 4.6 ([7]). *If A, B are definable subsets of G, $p \in S_1^G(M)$, $A \in p$, $B \subseteq \mathrm{St}_G(p)$, then there exists $g \in G$ such that $Bg \sqsubset A$.*

This implies

FACT 4.7 ([7, 43]). *G is connected if and only if G is irreducible, that is,* $D(G) = 1$.

FACT 4.8 ([7]). *If $\mathrm{St}_G(p) \in p$, then $\mathrm{St}_G(p)$ is connected.*

Cherlin [43] has studied groups G with $R(G) \le 3$.

FACT 4.9 ([43]). *If G is connected and $R(G) = 2$, then G is solvable. If at the same time the center of G is trivial, then G is isomorphic to a group of matrices of the form $\left(\begin{smallmatrix} a & b \\ 0 & 1 \end{smallmatrix}\right)$, where $a, b \in K$, $a \neq 0$, K being an algebraically closed field.*

COROLLARY 4.10. *Let G be a nonnilpotent connected group, $R(G) = 2$. Then one can define operations $+$ and $\cdot$ on the commutator subgroup U of G, definable in G, in such a way that $(U, +, \cdot)$ is an algebraically closed field.*

PROOF. Let Z be the center of G. In view of Fact 4.9, the center of G is finite. Since G is connected, any finite normal subgroup of G lies in the center, and therefore G/Z has no nontrivial finite normal subgroups. In particular, the center of G/Z is trivial, so that G/Z is isomorphic to a group $\{\left(\begin{smallmatrix} a & b \\ 0 & 1 \end{smallmatrix}\right) : a, b \in K,\ a \neq 0\}$, where K is an algebraically closed field. Denote the group G/Z by $\overline{G}$ and its elements by $\bar{g}$, if g is a representative of the class $\bar{g}$.

Obviously, $R(U) = 1$, and since G is connected so is U; that is, U is strongly minimal. If $u \in U - Z$, then $u^G = \{g^{-1} \cdot u \cdot g : g \in G\}$ is infinite, as $\bar{u}^{\overline{G}}$ is infinite. Then for any z in $U \cap Z$ we have $z \cdot u^G \cap u^G \neq \varnothing$. Hence $z \cdot u^{g_1} = u^{g_2}$ for some $g_1, g_2 \in G$. Going over to $\overline{G}$, we obtain $\bar{u}^{\bar{g}_1} = \bar{u}^{\bar{g}_2}$, and by the description of $\overline{G}$ we immediately obtain $\bar{g}_1 \cdot \overline{U} = \bar{g}_2 \cdot \overline{U}$, that is, $g_1 \cdot U \cdot Z = g_2 \cdot U \cdot Z$. Therefore, $u^{g_1} = u^{g_2}$, and so $z = 1$. We have thus proved that $U \cap Z = \{1\}$.

Since the commutator subgroup of $\overline{G}$ is isomorphic to $U/U \cap Z$, it follows that the commutator subgroup of $\overline{G}$ is isomorphic to U by a definable isomorphism. On the other hand, it follows from the description of $\overline{G}$ (Fact 4.9) that one can define operations $+$ and $\cdot$ on the commutator subgroup of $\overline{G}$, hence also on U; thus one can indeed define on the latter a structure isomorphic to the field K. □

§5. Totally categorical structures

A structure $\mathfrak{M}$ is said to be *totally categorical* if its theory is categorical in all infinite powers.

Thus, totally categorical structures are uncountably categorical structures which are also countably categorical.

The following fact is fundamental for the study of countably categorical theories.

FACT 5.1. *$\mathfrak{M}$ is countably categorical if and only if the Boolean algebras F_n are finite for all n.*

COROLLARY 5.2. *Let $\mathfrak{M}$ be countably categorical. Then any enrichment of $\mathfrak{M}$ by finitely many constants is countably categorical.*

$\mathfrak{M}$ is countably saturated.

If a structure $\mathfrak{N}$ is X-definable in $\mathfrak{M}$ and X is finite, then $\mathfrak{N}$ is countably categorical.

COROLLARY 5.3. *If $\mathfrak{M}$ is countably categorical, X is finite, and A is $\varnothing$-definable in $\mathfrak{M}$, then $A[X]$ is finite.*

FACT 5.4. *For a totally categorical structure $\mathfrak{M}$ and subsets X, A of $\mathfrak{M}$, the following conditions are equivalent:*

(a) *A is X-definable;*

(b) *A is stable under all automorphisms of $\mathfrak{M}$ whose restrictions to X are the identity.*

In the totally categorical case the polynomials appearing in the following theorem play a most important role, along with the Morley rank.

THEOREM 5.5. *Let A be a $\bar{z}$-definable set, C a $\varnothing$-definable strongly minimal set in a totally categorical structure $\mathfrak{M}$. If A is coordinatizable over $C_{\bar{z}}$, then there exist a polynomial $p_A(x)$ with rational coefficients and a natural number n_A such that for any finite X containing $|\bar{z}|$*

(a) *if $x = \|C[X]\| \geq n_A$, then $\|A[X]\| = p_A(x)$;*

(b) *$R(A) = \deg p_A$—the degree of p_A;*

(c) *if A is strictly k-coordinatizable over $C_{\bar{z}}$ for some k, then one can take $n_A = 0$;*

(d) *if*

$$A = \varphi(\bar{z}, \mathfrak{M}), \quad A' = \varphi(\bar{z}', \mathfrak{M}), \qquad \varphi \in F,$$

$$\operatorname{tp}(\bar{z}, \varnothing) = \operatorname{tp}(\bar{z}', \varnothing), \qquad |\bar{z}'| \subseteq \operatorname{acl}(X),$$

then $p_{A'}(x) = \|A'[X]\| = p_A(x)$. That is, the coefficients of p_A and the number n_A depend only on the type of $\bar{z}$.

PROOF. We may assume that A is a $\bar{z}$-atom. Let $a \in A$. Then there exists a tuple $\bar{c} = \langle c_1, \dots, c_k\rangle$ of elements of C such that $a \in \operatorname{acl}(|\bar{c}| \cup |\bar{z}|)$ and $c_1, \dots, c_k$ are algebraically independent over $|\bar{z}|$. Let ψ be a $\bar{z}$-atom such that $\mathfrak{M} \models \psi(\bar{c}, a)$.

Let $\overset{k}{C_Y^{\sim}}$ denote the set of all k-tuples of elements of C whose elements are algebraically independent over Y.

It is easy to see that $\psi(\bar{d}, \mathfrak{M}) \subseteq A[X]$ for any $\bar{d}$ in $\overset{k}{C_{\bar{z}}^{\sim}}[X]$; moreover $\|\psi(\bar{d}, \mathfrak{M})\| = r > 0$ and r is independent of the choice of $\bar{d}$. In addition,

$$\|\psi(\mathfrak{M}, a) \cap \operatorname{acl}(X)\| = u_X$$

is independent of the choice of a in A, since A is a $\bar{z}$-atom. Let

$$B = \{\bar{d}^\frown e : \mathfrak{M} \models \psi(\bar{d}, e)\}.$$

Counting the number of elements of $B[X]$ in two ways, we obtain

$$u_X \cdot \|A[X]\| = \|B[X]\| = \|\overset{k}{C_{\bar{z}}^{\sim}}[X]\| \cdot r.$$

It is easy to see that for $j < k$ the numbers

$$s_j = \|C[\{c_1, \dots, c_j\} \cup |\bar{z}|]\|$$

are independent of the choice of $\bar{c}$ in $\overset{k}{\smile}{C}_{\bar{z}}$ and, if $x = \|C[X]\|$, then

$$\|\overset{k}{\smile}{C}_{\bar{z}}[X]\| = (x - s_0) \cdots (x - s_{k-1}).$$

We may assume that $c_1, \dots, c_i$ are algebraically independent over $\{a\} \cup |\bar{z}|$, $\hat{c} = \langle c_1, \dots, c_i \rangle$, $\operatorname{acl}(|\bar{c}| \cup |\bar{z}|) = \operatorname{acl}(|\hat{c}| \cup \{a\} \cup |\bar{z}|)$ for some $i < k$. Denote

$$H = \{\bar{h} \in M^{k-i} : \mathfrak{M} \models \psi(\hat{c}^\frown \bar{h}, a)\}.$$

It is clear that

$$H[X] \subseteq \operatorname{acl}(|\hat{c}| \cup \{a\} \cup |\bar{z}|),$$

and moreover $\|H[X]\| = q > 0$, and q is independent of the choice of a and $\bar{c}$.

The numbers $t_j = \|C[\{d_1, \dots, d_j, a\} \cup |\bar{z}|]\|$ for $j < i$ are independent of the choice of a and $\langle d_1, \dots, d_j \rangle$ in $\overset{j}{\smile}{C}_{\{a\}\cup|\bar{z}|}$ and, if $x = \|C[X]\|$, then

$$\|\overset{i}{\smile}{C}_{\{a\}\cup|\bar{z}|}[X]\| = (x - t_0) \cdots (x - t_{i-1}).$$

Obviously, if for $j \le i$

$$\dim_{\bar{z}} C[\{c_1, \dots, c_j, a\} \cup |\bar{z}|] = m(j),$$

then $m(j) \le k$ and $t_j = s_{m(j)}$. It is also easy to see that for $x = \|C[X]\|$:

$$u_X = q \cdot (x - t_0) \cdots (x - t_{i-1}).$$

Then if $x = \|C[X]\| > t_{i-1}$, we have $u_X > 0$ and

$$\|A[X]\| = \frac{\|\overset{k}{\smile}{C}_{\bar{z}}[X]\| \cdot r}{u_X} = \frac{r \cdot (x - s_0) \cdots (x - s_{k-1})}{q \cdot (x - t_0) \cdots (x - t_{i-1})}.$$

After canceling the fractions by all $x - t_j$, we obtain a polynomial $p_A(x)$ of degree $k - i$, and we set $n_A = t_{i-1} + 1$.

Since

$$R(a, |\bar{c}| \cup |\bar{z}|) + R(\bar{c}, |\bar{z}|) = R(\bar{c}, \{a\} \cup |\bar{z}|) + R(a, |\bar{z}|)$$

and

$$R(a, |\bar{c}| \cup |\bar{z}|) = 0, \quad R(\bar{c}, |\bar{z}|) = k, \quad R(\bar{c}, \{a\} \cup |\bar{z}|) = i,$$

it follows that

$$R(A) = R(a, |\bar{z}|) = k - i = \deg p_A.$$

If it were true that $\operatorname{acl}(\{a\} \cup |\bar{z}|) = \operatorname{acl}(|\bar{c}| \cup |\bar{z}|)$, then $i = 0$, $u_X = q$, and one can put $n_A = 0$.

Finally, as is evident from the construction, $r, q, s_0, \dots, s_{k-1}, t_0, \dots, t_{i-1}$, and hence also all the coefficients of p_A and n_A, depend only on the type of $\bar{z}$. $\square$

CHAPTER II

Three Types of Uncountably Categorical Structures

In this chapter we proceed to study the structure of strongly minimal structures definable in a fixed uncountably categorical structure $\mathfrak{M}$. In contrast to the global properties of $\mathfrak{M}$ (the structure of $\mathfrak{M}$ as a whole), properties of strongly minimal structures in $\mathfrak{M}$ are naturally referred to as local properties of $\mathfrak{M}$. The term is also natural in connection with the fact that every infinite definable set has a strongly minimal subset. As will be shown, in particular, in this chapter, the main local properties of $\mathfrak{M}$ are independent of the specific strongly minimal set under consideration in $\mathfrak{M}$, that is, "locally, all sets definable in $\mathfrak{M}$ are identically structured".

The main result of the chapter is the Trichotomy Theorem, which partitions the class of all uncountably categorical structures, in accordance with their properties, into three types: field-like, module-like and of disintegrated type. Essentially, these three types of structures require different approaches to the study of their global properties. In particular, it will be proved in §4 of this chapter that groups definable in a structure $\mathfrak{M}$ may be other than almost abelian only if $\mathfrak{M}$ is field-like; in structures of discrete type infinite groups are not definable at all. These results will be utilized in Chapter 5.

§1. Combinatorial definitions and preliminary results

A set A with a closure operator $c\colon 2^A \to 2^A$ is known as a *pregeometry* [2] if, for any subsets X, Y of A:

(1) $X \subseteq c(X)$;

(2) $c(X) - \bigcup\{c(X') : X' \subseteq X,\ X' \text{ finite}\}$;

(3) $X \subseteq c(Y)$ implies $c(X) \subseteq c(Y)$;

(4) for any $x, y \in A$, if $x \in c(X \cup \{y\}) - c(X)$, then $y \in c(X \cup \{x\})$.

A pregeometry is said to be a *geometry* if

(5) $c(\{x\}) = \{x\}$ for all $x \in A$ and $c(\varnothing) = \varnothing$.

A geometry is said to be *degenerate* if $c(X) = X$ for all $X \subseteq A$.

A pregeometry is said to be *locally finite* if $c(X)$ is finite for all finite subsets X.

A pregeometry is said to be *homogeneous* if, for any $X \subseteq A$ and any $y_1, y_2 \in A - c(X)$, there is an automorphism of the pregeometry (i.e., a map preserving c) that takes y_1 onto y_2 and is the identity on X.

Closed subsets of A are called *subspaces*. A subset X of a subspace S is called a *basis* of S if $c(X) = S$ and X is a minimal subset with this property. The *dimension* of S is the cardinality of a basis; it is denoted by $\dim S$. It follows easily from the pregeometry axioms that dimension is well defined. When one is speaking of geometries, the two-dimensional subspaces are called the *lines* and the three-dimensional subspaces the *planes* of the geometry.

To any pregeometry A one *associates*, in a natural way, *a geometry* $\widehat{A}$ whose points are the sets $c(\{x\})$, where $x \in A - c(\varnothing)$, with the closure of a subset

$$\widehat{X} = \{c(\{x\}) : x \in X - c(\varnothing)\} \subseteq \widehat{A}$$

defined as the set

$$c(\widehat{X}) = \{c(\{y\}) : y \in c(X) - c(\varnothing)\}.$$

On the other hand, given a pregeometry A and an arbitrary subset Y of A, one can define a pregeometry A_Y with the same set of points and the new closure operator

$$c_Y(X) = c(Y \cup X).$$

The geometry $\widehat{A_Y}$ is called the *localization of* A *with respect to* Y.

PROPOSITION 1.1. (a) *If A is a strongly minimal set in a structure $\mathfrak{M}$, then A together with the operator* acl *is a homogeneous pregeometry. If moreover $\mathfrak{M}$ is also countably categorical, the pregeometry $\langle A, \text{acl}\rangle$ is locally finite.*

(b) *Let $\langle A, c\rangle$ be an infinite homogeneous locally finite pregeometry. Consider the structure $\mathfrak{A} = \langle A, C^{(0)}, \dots, C^{(n)}, \dots\rangle$, where $C^{(n)}(x_1, \dots, x_n, y)$ is the $(n+1)$ place predicate on A meaning that $y \in c(\{x_1, \dots, x_n\})$. Then this structure is countably categorical and strongly minimal, and the operator c coincides with the operator* acl *in $\mathfrak{A}$.*

PROOF. (a) was pointed out by Marsh and was proved in [38 (Chapter 1, §2)].

(b) Fix an arbitrary finite $X \subseteq A$ and consider the complete 1-types over X realizable in $\mathfrak{A}$. By homogeneity, all elements of $A - c(X)$ realize the same type over X, and since $c(X)$ is finite there are only finitely many complete 1-types realizable in $\mathfrak{A}$. Hence it follows that the set of all complete 1-types over X is finite, since the set of realizable complete types is dense in the set of all complete types. Hence, by the Ryll-Nardzewski Theorem, $\mathfrak{A}$ is countably categorical. These arguments also imply incidentally that $c(X) = \text{acl}(X)$ for any X.

Since $\mathfrak{A}$ is countably categorical, $\mathfrak{A}$ is countably saturated. We have already remarked that A cannot be split into two infinite X-definable sets. Since the structure is countably saturated, this implies that A is strongly minimal. □

Proposition 1.1(a) associates with any strongly minimal set A a pregeometry $\langle A, \mathrm{acl}\rangle$. The corresponding geometry $\widehat{A}$ will be called the *geometry of the strongly minimal set* A.

Proposition 1.1.(b) makes it possible to treat infinite locally finite homogeneous pregeometries as strongly minimal structures, as we shall do repeatedly in the sequel.

Let V be a vector space over a division ring K. Defining $c(X)$, for any $X \subseteq V$, as the subspace spanned by X, we obtain a homogeneous pregeometry. If K is finite, this pregeometry is locally finite.

A geometry isomorphic to $\widehat{V}$ for some vector space V is called a *projective geometry*. A geometry A is said to be *locally projective* if its localization $\widehat{A}_a$ with respect to any point $a \in A$ is projective.

One can also associate another homogeneous geometry with a given vector space V. Define $c(X)$ to be the smallest subset of V which, together with every pair of distinct points v_1, v_2, contains the points $v_1 + k(v_2 - v_1)$ for all $k \in K$ (the line passing through v_1, v_2). A geometry isomorphic to this geometry is called an *affine geometry*.

It is easy to see that an affine geometry is locally projective.

The following partial converse to this fact was established by Doyen and Hubaut [47].

FACT 1.2. *Any locally projective finite geometry of dimension at least four, all lines of which have the same cardinality, is an affine or projective geometry over a finite field. In particular, every locally projective infinite locally finite homogeneous geometry is an affine or projective geometry over a finite field.*

It is well known ([77, 46, 2]) that affine and projective geometries of dimension greater than two may be characterized in terms of relations between subspaces, hence in terms of closure. In particular, a geometry A is projective if and only if every line contains at least three points and

(6) for all $y \in A$, $Z \subseteq A$, $x \in c(\{y\} \cup Z)$, there exists $z \in c(Z)$ such that $x \in c(\{y, z\})$.

Affine geometries are characterized as geometries obtained by removing all points of a subspace of codimension 1 from a projective geometry.

An *incidence system* is defined as a triple (P, L, I), where P and L are sets of "points" and "lines", respectively, and $I \subseteq P \times L$ is a relation, called incidence. An incidence system may be considered as a structure $\mathscr{P}$ with underlying set $P \cup L$, one-place predicates P, L and a two-place predicate I. It is in this sense that we shall understand the definability of an incidence system $\mathscr{P}$ in other structures.

Given a point $p_0 \in P$ and a line $l_0 \in L$, we denote

$$p_0 I = \{l \in L : p_0 I l\},$$
$$I l_0 = \{p \in P : p I l_0\}.$$

A *pseudoplane* is an incidence system (P, L, I) which satisfies the following conditions:

(7) pI is infinite for every $p \in P$;

(8) Il is infinite for every $l \in L$;

(9) $p_1 I \cap p_2 I$ is at most finite for different $p_1, p_2 \in P$;

(10) $Il_1 \cap Il_2$ is at most finite for different l_1, l_2.

§2. Pseudoplanes in strongly minimal structures

The aim of this section is to construct a pseudoplane which is definable in $\mathfrak{A}$, where $\mathfrak{A}$ is a strongly minimal structure which is definable in $\mathfrak{M}$ and satisfies certain additional conditions.

LEMMA 2.1. *Let (P, L, I) be an incidence system which is $\varnothing$-definable in $\mathfrak{M}$,*

$$p_0 \in P, \quad l_0 \in L, \quad \langle p_0, l_0 \rangle = I,$$
$$R(l_0, \varnothing) = R(L), \qquad R(p_0, \varnothing) = R(P),$$
$$R(p_0, \{l_0\}) = R(Il_0), \qquad R(l_0, \{p_0\}) = R(p_0 I).$$

Then there exist a system (P', L', I') which is $\varnothing$-definable in $\mathfrak{M}$, a line $l'_0 \in L'$ and a $\varnothing$-definable map $\pi\colon L' \to L$ such that $P' = P$, $R(L') = R(L)$, $\pi^{-1}(l)$ is finite for all $l \in L$,

$$\pi(l'_0) = l_0, \quad R(I'l'_0) = R(Il_0), \quad \mathrm{D}(I'l'_0) = 1,$$

for all $p \in P$, $\pi(pI') \subseteq pI$.

PROOF. By the Finite Equivalence Relation Theorem (Fact 1.3.6), there exists a formula E_{l_0}, with two variables and a constant l_0, which defines an equivalence relation on Il_0 with finitely many classes, each class of which is of degree 1 or of rank less than $r_0 = R(Il_0)$. Let k_0 be the number of classes.

Since the union of all classes of rank less than r_0 is l_0-definable and of rank less than r_0, it follows that p_0 lies in a class of rank r_0 and degree 1.

Define

$$L_1 = \{l \in L : E_l \text{ is an equivalence relation on } Il \text{ with } k_0 \text{ classes}\}.$$

Clearly, L_1 is $\varnothing$-definable, $l_0 \in L_1$; therefore $R(L_1) = R(L)$.

Define an equivalence relation E on $I \cap (P \times L_1)$ as follows:

$$\langle p_1, l_1 \rangle E \langle p_2, l_2 \rangle \Leftrightarrow l_1 = l_2 \;\&\; p_1 E_{l_1} p_2,$$

and set

$$L' = I \cap (P \times L_1)/E.$$

It is easy to see that for any $l \in L_1$ there exist exactly k_0 elements $l' \in L'$ such that $l' = \langle p, l \rangle E$ and l' uniquely determines l. Define $\pi(l') = l$ in this case. Obviously, $\operatorname{acl}(l') = \operatorname{acl}(l)$. Therefore, in particular, $R(L') = R(L)$.

Define

$$pI'l' \Leftrightarrow l' = \langle p, l \rangle E \;\&\; R(\langle p, l \rangle E) = r_0,$$

and

$$l_0' = \langle p_0, l_0 \rangle E.$$

Obviously, $I'l_0' \subseteq Il_0$, $I'l_0'$ is the equivalence class modulo E_{l_0} containing p_0, and therefore $R(I'l_0') = r_0$, $D(I'l_0') = 1$.

Finally, the last condition follows directly from the definition of π and I'. □

LEMMA 2.2. *If A contains elements a_1, a_2, b_1, b_2, c, every four of which are algebraically independent, $c \in \mathrm{acl}(a_1, a_2, b_1, b_2)$ and $\mathrm{acl}(a_1, a_2, c) \cap \mathrm{acl}(b_1, b_2, c) = \mathrm{acl}(c)$, then an incidence system (P, L, I) such that*

$\mathrm{RD}(P) = \langle 2, 1 \rangle$, $R(L) \geq 2$;
$R(Il) = 1$ *for all* $l \in L$;
if $l_1, l_2 \in L$, $l_1 \neq l_2$, *then* $R(Il_1 \cap Il_2) = 0$ *is* $\varnothing$*-definable in* $\mathfrak{A}$.

PROOF. Let $P_0 = A \times A$, $L_0 = A \times A \times A$, and let $I_0 \subseteq P_0 \times L_0$ be an arbitrary $\varnothing$-definable relation such that

$$\langle b_1, b_2 \rangle I_0 \langle a_1, a_2, c \rangle,$$

$$\langle x_1, x_2 \rangle I_0 \langle y_1, y_2, z \rangle \to z \in \mathrm{acl}(x_1, x_2, y_1, y_2).$$

It is easy to verify that, once we denote $p_0 = \langle b_1, b_2 \rangle$, $l_0 = \langle a_1, a_2, c \rangle$, we are in the situation of Lemma 2.1. Consequently, for some L_0', $l_0' \in L_0'$, I_0' we have

$$R(L_0') = R(L_0) = 3, \qquad \mathrm{acl}(l_0') = \mathrm{acl}(l_0),$$

and $I_0'l_0'$ is strongly minimal.

Define

$$L_1 = \{l_1 \in L_0' : R(I_0'l_1) = 1 \;\&\; (\forall l_2 \in L_0')(R(I_0'l_1 \cap I_0'l_2) > 0 \to I_0'l_1 \sqsubset I_0'l_2)\}.$$

Since $I_0'l_0'$ is strongly minimal, it follows that $l_0' \in L_1$, so that

$$R(L_1) = R(l_0', \varnothing) = R(l_0, \varnothing) = 3.$$

Define a $\varnothing$-definable equivalence relation E on L_1 by

$$l_1 E l_2 \Leftrightarrow I_0'l_1 \sqsubset\!\sqsupset I_0'l_2.$$

Now put $P = P_0$, $L = L_1/E$, and for $p \in P$, $l \in L_1$

$$pI(lE) \Leftrightarrow pI_0' \sqsupset lE.$$

We now apply Proposition 1.2.7, setting $\gamma(\mathfrak{M}) = I_0'$, $A = lE$, B an arbitrary l-definable infinite subset of $I_0'l$. For any finite $X \subseteq lE$, the set

$$I_0'X = \bigcap\{I_0'x : x \in X\},$$

as follows from the definition of E, almost coincides with $I_0'l$. Therefore, since $R(I_0'l) = 1$ $(l \in L_1)$ and B is finite, there exists $p \in I_0'X \cap B$. In other words, there exists $p \in B$ such that $pI_0' \supseteq X$. Now, by Proposition 1.2.7, we can also find a point $p \in B$ such that $pI'_0 \sqsupset lE$. Finally, since such a point may be found in any infinite l-definable subset B of $I'_0 l$, it follows that $I(lE) \sqsupset I'_0 l$, whence $R(I(lE)) = 1$. In particular, for our p_0 and l'_0, putting $\bar{l}_0 = l'_0 E$, we have: $I\bar{l}_0$ is strongly minimal and $p_0 I \bar{l}_0$.

If $l_1, l_2 \in L_1$, $\bar{l}_1 = l_1 E$, $\bar{l}_2 = l_2 E$, $R(I\bar{l}_1 \cap I\bar{l}_2) > 0$, then $R(I'_0 l_1 \cap I'_0 l_2) > 0$, and therefore $I'_0 l_1 \sqsupset I'_0 l_2$ by the definition of L_1. Therefore, $\bar{l}_1 = \bar{l}_2$.

We now show that $R(\bar{l}_0, \varnothing) \geq 2$, and so $R(L) \geq 2$.

Suppose that $R(\bar{l}_0, \varnothing) \leq 1$. Then, since $R(p_0, \{\bar{l}_0\}) = 1 < R(p_0, \varnothing)$, we have $R(\bar{l}_0, \{p_0\}) < R(\bar{l}_0, \varnothing) \leq 1$, i.e., $\bar{l}_0 \in \operatorname{acl}(p_0)$. Since $p_0 = \langle b_1, b_2 \rangle$, it follows that $\bar{l}_0 \in \operatorname{acl}(b_1, b_2)$. Therefore, obviously, $c \notin \operatorname{acl}(\bar{l}_0)$. On the other hand, $b_1 \notin \operatorname{acl}(l_0)$, since $\bar{l}_0 \in \operatorname{acl}(l_0) = \operatorname{acl}(a_1, a_2, c)$. Therefore there exists $c' \in A$ such that $\operatorname{tp}(\langle c, c' \rangle, \{\bar{l}_0\}) = \operatorname{tp}(\langle b_1, b_2 \rangle, \{\bar{l}_0\})$. Since $R(\langle c, c' \rangle, \{\bar{l}_0\}) = 1$, it follows that $c' \in \operatorname{acl}(\bar{l}_0, c) \subseteq \operatorname{acl}(b_1, b_2, c) \cap \operatorname{acl}(a_1, a_2, c)$. By definition, $c' \in \operatorname{acl}(c)$. But this contradicts the assumption of the lemma. Consequently, $R(\bar{l}_0, \varnothing) \geq 2$. □

LEMMA 2.3. *Let (P, L, I) be an incidence system which is $\varnothing$-definable in $\mathfrak{M}$, $p_0 \in P$, $R(p_0, \varnothing) \geq R(L)$, $R(p_0 I) > 0$, and assume that if $p_1, p_2 \in P$, $p_1 \neq p_2$, then $R(p_1 I \cap p_2 I) = 0$. Then there exists l_0 in L such that*

$$\langle p_0, l_0 \rangle \in I,\ R(p_0, \{l_0\}) > 0,$$

$$R(Il_0) > 0,\ R(l_0, \{p_0\}) = R(p_0 I),\ R(l_0, \varnothing) > R(p_0 I).$$

PROOF. Suppose there exists a $\varnothing$-definable $L' \subseteq L$ such that $R(L') = R(p_0 I \cap L') > 0$. Then the condition $R(p_1 I \cap p_2 I) = 0$ implies that there exist at most finitely many elements p in P such that $R(pI \cap L') = R(L')$— contradiction, since p may be any point such that $\operatorname{tp}(p, \varnothing) = \operatorname{tp}(p_0, \varnothing)$, and by assumption $R(p_0, \varnothing) \geq R(L) \geq R(p_0 I) > 0$, i.e., there are infinitely many such points p.

Thus, for any $\varnothing$-definable $L' \subseteq L$ intersecting $p_0 I$, we have $R(L') > R(p_0 I \cap L')$ if $R(L') > 0$. Hence it follows that $R(l_0, \varnothing) > R(p_0 I)$ for any $l_0 \in p_0 I$ such that $R(l_0, \{p_0\}) = R(p_0 I)$. Evaluating

$$R(\langle p_0, l_0 \rangle, \varnothing) = R(p_0, \{l_0\}) + R(l_0, \varnothing) = R(l_0, \{p_0\}) + R(p_0, \varnothing),$$

we obtain

$$R(p_0, \{l_0\}) = R(l_0, \{p_0\}) + R(p_0, \varnothing) - R(l_0, \varnothing) \geq R(l_0, \{p_0\}) > 0,$$

and therefore $R(Il_0) > 0$. □

LEMMA 2.4. *Let A contain elements a_1, a_2, b_1, b_2, c, every four of which are algebraically independent, $c \in \operatorname{acl}(a_1, a_2, b_1, b_2)$ and*

$$\operatorname{acl}(a_1, a_2, c) \cap \operatorname{acl}(b_1, b_2, c) = \operatorname{acl}(c).$$

Then there is a pseudoplane (P', L', I') definable in $\mathfrak{A}$ such that

$$\operatorname{RD}(P') = \operatorname{RD}(L') = \langle 2, 1 \rangle.$$

PROOF. We use Lemma 2.2. In view of the symmetry of the definition, this lemma yields an incidence system (P_0, L_0, I_0), definable in $\mathfrak{A}$, such that

(i) $R(P_0) \geq 2$, $\mathrm{RD}(L_0) = \langle 2, 1\rangle$;
(ii) $R(pI_0) = 1$ for all $p \in P_0$;
(iii) if $p_1, p_2 \in P_0$, $p_1 \neq p_2$, then $R(p_1 I_0 \cap p_2 I_0) = 0$.

By considering, instead of P_0, a subset of it of rank 2 and degree 1 and expanding the structure with constants, we preserve all the conditions. We may, therefore, assume that $\mathrm{RD}(P_0) = \langle 2, 1\rangle$ and that (P_0, L_0, I_0) is $\varnothing$-definable.

Choose $p_0 \in P$ so that $R(p_0, \varnothing) \geq 2$, and using Lemma 2.3 find $l_0 \in L_0$ such that $\langle p_0, l_0\rangle \in I_0$, $R(p_0, \varnothing) = R(l_0, \varnothing) = 2$, $R(p_0 I) = R(l_0, \{p_0\}) = 1$, $R(I_0 l_0) = R(p_0, \{l_0\}) = 1$.

Now, using Lemma 2.1, we go over to an incidence system (P_0, L_0', L_0'). In this system $I_0' l_0'$ is strongly minimal for some l_0' such that $R(l_0', \varnothing) = 2$. In addition, if $p_1, p_2 \in P$ are distinct, the intersection $p_1 I_0' \cap p_2 I_0'$ is finite, since

$$p_1 I_0' \cap p_2 I_0' \subseteq \pi^{-1}(p_1 I_0 \cap p_2 I_0).$$

Now, as in the proof of Lemma 2.2, define

$$L_1 = \{l_1 \in L_0' : R(I_0' l_1) = 1 \;\&\; (\forall l_2 \in L_0')(R(I_0' l_1 \cap I_0' l_2) > 0 \to I_0' l_1 \sqsubset I_0' l_2)\},$$
$$l_1 E l_2 \Leftrightarrow I_0' l_1 \sqsubset\!\!\!\sqsupset I_0' l_2,$$
$$P = P_0, \qquad L = L_1/E,$$
$$pI(lE) \Leftrightarrow pI_0' \sqsupset lE, \qquad \bar{l}_0 = l_0' E.$$

As in Lemma 2.2, we obtain $p_0 I \bar{l}_0$, $R(I\bar{l}) = 1$ for all $\bar{l} \in L$, $R(I\bar{l}_1 \cap I\bar{l}_2) = 0$ for distinct $\bar{l}_1, \bar{l}_2 \in L$.

Note that each class lE is finite. Indeed, suppose that lEl_1 is valid. Then there exist $p_1, p_2 \in I_0' l_1 \cap I_0' l$ such that

$$R(p_1, \{l\}) = R(p_1, \{l, l_1\}) = 1, \qquad R(p_2, \{p_1, l\}) = R(p_2, \{p_1, l, l_1\}) = 1.$$

Then, by the Reciprocity Principle,

$$R(l_1, \{l\}) = R(l_1, \{l, p_1\}), \qquad R(l_1, \{l, p_1\}) = R(l_1, \{l, p_1, p_2\}),$$

that is,

$$R(l_1, \{l\}) \leq R(l_1, \{p_1, p_2\}) = 0.$$

This follows from the fact that $l_1 \in p_1 I_0' \cap p_2 I_0'$.

Since lE is finite, $pI(lE)$ is equivalent to $pI_0' \supseteq lE$, and so $p_1 I \cap p_2 I \subseteq p_1 I_0' \cap p_2 I_0'/E$. Hence $R(p_1 I \cap p_2 I) = 0$ if $p_1 \neq p_2$. But since lE is finite for all $l \in L_1$, it follows that $\mathrm{RD}(L) = \mathrm{RD}(L_1) = \langle 2, 1\rangle$. In addition, $R(\bar{l}_0, \{p_0\}) = R(l_0, \{p_0\})$, since $\mathrm{acl}(\bar{l}_0) = \mathrm{acl}(l_0)$. Hence $R(p_0 I) = 1$.

Let $P' = \{p \in P : R(pI) = 1\}$. Since $p_0 \in P'$, we have $\mathrm{RD}(P') = \langle 2, 1\rangle$, $R(P - P') \leq 1$. Define

$$L' = \{l \in L : R(Il \cap P') = 1\}.$$

Since $R(P - P') \leq 1$, the difference $L - L'$ is at most finite. Now, defining $I' = I \cap (P' \times L')$, we obtain a pseudoplane (P', L', I'). □

PROPOSITION 2.5. *If $\mathfrak{A}$ is a strongly minimal structure in which no pseudoplane (P, L, I) with $\mathrm{RD}(P) = \mathrm{RD}(L) = \langle 2, 1\rangle$ is definable, then the geometry $\widehat{A}$ is locally projective or degenerate.*

PROOF. By definition, to say that $\widehat{A}$ is locally projective means that for any $c \in A - \mathrm{acl}(\varnothing)$:

(*) $\mathrm{acl}(x, y, c) \neq \mathrm{acl}(x, c) \cup \mathrm{acl}(y, c)$ if $x, y \in A$, and x, y, c are independent;

(**) if $y \in A$, $Z \subseteq A$, $x \in \mathrm{acl}(\{y, c\} \cup Z)$, there exists $z \in \mathrm{acl}(c, Z)$ such that $x \in \mathrm{acl}(y, z, c)$.

We first show that if no pseudoplane is definable in $\mathfrak{A}$, then condition (**) holds. Assuming Z to be finite, we prove (**) by induction on the number of elements in Z.

If $Z = \{z\}$, there is nothing to prove. Let $Z = \{z_1, z_2\} \cup Z_0$, $x \in \mathrm{acl}(\{y, c\} \cup Z)$. We may assume that y, z_1, z_2, c are independent over Z_0, for otherwise y, z_1, or z_2 could be omitted. Adding Z_0 to the signature, we use Lemma 2.4. Then either any four elements out of x, y, z_1, z_2, c are algebraically dependent over Z_0, or there exists $z \in A - A[\{c\} \cup Z_0]$,

$$\mathrm{acl}(\{c\} \cup Z) \cap \mathrm{acl}(\{x, y, c\} \cup Z_0) \supseteq \mathrm{acl}(\{z, c\} \cup Z_0).$$

In the first case, the only possibility is $x \in \mathrm{acl}(\{y, z_1, z_2\} \cup Z_0)$. Since $z_1 \in A - \mathrm{acl}(\varnothing)$ and $\|\{z_2\} \cup Z_0\| < \|Z\|$, it follows from the inductive hypothesis (with z_1 instead of c) that there exists $z \in \mathrm{acl}(Z)$ such that $x \in \mathrm{acl}(y, z, z_1)$. If $x \in \mathrm{acl}(y, z, c)$ or $x \in \mathrm{acl}(y, z_1, c)$, we obtain the desired assertion. Otherwise, $\dim A[y, z_1, z, c] = 4$ and we can choose $z_1', z_2' \in A[z_1, z, c]$ such that any quadruple from $\{x, y, z_1', z_2', c\}$ is independent. By 2.4 we can find $z' \in A[z_1', z, c] \cap A[x, y, c] - A[c]$. Clearly,

$$z' \in \mathrm{acl}(\{c\} \cup Z), \qquad x \in \mathrm{acl}(y, z', c).$$

In the second case we have $z \in \mathrm{acl}(\{c\} \cup Z)$, $z \in \mathrm{acl}(\{x, y, z\} \cup Z_0)$. At the same time, $z \notin \mathrm{acl}(\{y, c\} \cup Z_0)$, since otherwise $y \in \mathrm{acl}(\{z, c\} \cup Z_0) \subseteq \mathrm{acl}(\{c\} \cup Z)$, and we have assumed that y, z_1, z_2, c are independent over Z_0. Hence it follows that $x \in \mathrm{acl}(\{y, z, c\} \cup Z_0)$. Since $\|\{z\} \cup Z_0\| < \|Z\|$, the inductive hypothesis yields the existence of $z' \in \mathrm{acl}(\{z, c\} \cup Z_0) \subseteq \mathrm{acl}(\{c\} \cup Z)$ such that $x \in \mathrm{acl}(y, z', c)$, completing the proof of (**).

Now, if (*) is valid, we deduce that $\widehat{A}$ is locally projective. Otherwise, $\mathrm{acl}(x_0, x_1) \cup \mathrm{acl}(x_0, x_2) = \mathrm{acl}(x_0, x_1, x_2)$, if $x_0 \in A - \mathrm{acl}(x_1, x_2)$.

We claim that then $\mathrm{acl}(x_1, x_2) = \mathrm{acl}(x_1) \cup \mathrm{acl}(x_2)$. Let us assume that $x_0 \notin \mathrm{acl}(x_1, x_2)$, and let

$$y \in \mathrm{acl}(x_1, x_2) - (\mathrm{acl}(x_1) \cup \mathrm{acl}(x_2)).$$

Then $x_0 \in \mathrm{acl}(x_1, y) = \mathrm{acl}(x_1, x_2)$ or $x_0 \in \mathrm{acl}(x_2, y) = \mathrm{acl}(x_1, x_2)$, since $y \in \mathrm{acl}(x_0, x_1) \cup \mathrm{acl}(x_0, x_2)$, contradiction.

Now let $X = \{x_0, x_1, \dots, x_n\}$. We shall prove by induction on n that

$$\mathrm{acl}(X) = \mathrm{acl}(x_0, x_1) \cup \cdots \cup \mathrm{acl}(x_0, x_n).$$

Indeed, $\mathrm{acl}(X)$ obviously contains the right-hand side of this equality. If now $y \in \mathrm{acl}(x_0, x_1, \dots, x_n)$, then by $(**)$ there exists $x \in \mathrm{acl}(x_0, x_1, \dots, x_{n-1})$ such that $y \in \mathrm{acl}(x, x_n, x_0) = \mathrm{acl}(x_0, x_n) \cup \mathrm{acl}(x_0, x)$. But $\mathrm{acl}(x_0, x) \subseteq \mathrm{acl}(x_0, x_1) \cup \cdots \cup \mathrm{acl}(x_0, x_{n-1})$ by the inductive hypothesis, and this implies that $\mathrm{acl}(X)$ is a subset of the right-hand side and proves the equality.

It was proved about that $\mathrm{acl}(x_0, x_i) = \mathrm{acl}(x_0) \cup \mathrm{acl}(x_i)$. Hence

$$\mathrm{acl}(x_0, x_1, \dots, x_n) = \mathrm{acl}(x_0) \cup \cdots \cup \mathrm{acl}(x_n),$$

so that $\widehat{A}$ is degenerate. □

PROPOSITION 2.6. *Let* $\mathscr{P} = (P, L, I)$ *be a pseudoplane such that* $\mathrm{RD}(P) = \langle 2, 1\rangle$, $\mathrm{RD}(L) = \langle 2, 1\rangle$. *Then* $\mathscr{P}$ *is an uncountably categorical structure.*

PROOF. Since $R(L) < \infty$, we can find a strongly minimal subset N in L. We claim that if $\|N\| = \lambda$, then $\|P\| = \|L\| = \lambda$. If that is the case then, by the Baldwin-Lachlan Criterion (Fact 1.1.6), the theory of the structure $\mathscr{P}$ is uncountably categorical.

Choose an arbitrary strongly minimal $K \subseteq L$. Let IN, IK be the sets of points of P on the lines of N and K, respectively. It is easy to see that $R(IN) \geq 2$, $R(IK) \geq 2$. Since $\mathrm{RD}(P) = \langle 2, 1\rangle$, it follows that $IN \sqsubset P$, $IK \sqsubset P$, $IN \cap IK \sqsubset P$, $R(P - IN \cap IK) \leq 1$.

There is an infinite set of points on each line in L, and moreover, different lines intersect in at most a finite set. Hence there exists a line $l \notin N \cup K$ such that $Il \cap (P - IN \cap IK)$ is either finite or empty, and consequently $Il \cap IN \cap IK$ is infinite.

Since N is strongly minimal, it follows that for all but a finite number of lines $n \in N$ there exists a finite number of lines k in K such that $Il \cap In \cap Ik \neq \varnothing$. Hence $\|N\| \leq \|K\|$. By symmetry, $\|N\| = \|K\| = \lambda$.

We have thus shown that the cardinality of any subset of L of rank 1 is λ, since any such subset is the disjoint union of finitely many strongly minimal subsets.

Let U be a strongly minimal subset of P, $u \in U$,

$$P_u = \bigcup\{Il : l \in uI\},$$

that is, P_u is the set of points connected to u by a line. Since uI is infinite, $R(P_u) = 2$, $P_u \sqsubset P$. Therefore $P_{u_1} \cap \cdots \cap P_{u_n} \neq \varnothing$ for any $u_1, \dots, u_n \in U$.

Hence it follows that there exists a point $p \in P$ which can be connected to almost all points of U. Let L_U^p denote the set of lines passing through p and intersecting U at finitely many points. It follows from our previous remarks that $\|L_U^p\| = \|U\|$ and $R(L_U^p) = R(U) = 1$. Consequently, $\|U\| = \lambda$. We have thus shown that any definable subset of P of rank q has cardinality λ.

Note that $R(Il) = 1$, and therefore $\|Il\| = \lambda$ for all lines l except perhaps one "pathological" line. We may assume that N does not contain the pathological line. Then the cardinality of IN is λ. But $R(IN) = 2$, $R(P - IN) \leq 1$, and consequently $\|P - IN\| \leq \lambda$, $\|P\| = \lambda$. We now immediately obtain $\|L\| = \lambda$, since otherwise there is a pair of distinct points in P through which infinitely many lines pass. □

COROLLARY. *The pseudoplane* $(P, L, I) = \mathscr{P}$ *in Proposition 2.5 is uncountably categorical.*

Indeed, under the interpretation of $\mathscr{P}$ in $\mathfrak{A}$, $\mathrm{RD}(P) = \mathrm{RD}(L) = \langle 2, 1\rangle$, and so $\mathrm{RD}(P)$ and $\mathrm{RD}(L)$, evaluated in the structure $\mathscr{P}$, do not exceed $\langle 2, 1\rangle$. On the other hand, since there are infinitely many infinite lines in $\mathscr{P}$, it follows that $R(P) = R(L) = 2$ in $\mathscr{P}$. Thus, $\mathrm{RD}(P) = \mathrm{RD}(L) = \langle 2, 1\rangle$ in $\mathscr{P}$. The rest follows from Proposition 2.6.

§3. The Trichotomy Theorem

PROPOSITION 3.1. *If a pseudoplane is definable in* $\mathfrak{M}$, $X \subseteq M$, *then for any strongly minimal structure* $\mathfrak{A}$ *which is* X*-definable in* $\mathfrak{M}$ *the geometry* $\widehat{A}_X$ *associated with* $\mathfrak{A}$ *over* X *is neither degenerate nor projective.*

PROOF. Let $X' \supseteq X$ and let (P, L, I) be an incidence system, X'-definable in $\mathfrak{M}$, which satisfies the following conditions:

(1) $R(Il_1 \cap Il_2) = 0$, $R(p_1 I \cap p_2 I) = 0$, for $l_1, l_2 \in L$, $p_1, p_2 \in P$, $l_1 \neq l_2$, $p_1 \neq p_2$;

(2) for some pair $\langle p_0, l_0\rangle \in I$,

$$R(p_0, X') = R(P), \quad R(l_0, X') = R(L),$$
$$R(l_0, \{p_0\} \cup X') > 0, \quad R(Il_0) > 0, \quad R(p_0 I) > 0;$$

(3) the quadruple $\langle R(P), D(P), R(L), D(L)\rangle$ is minimal in the lexicographical ordering of all such quadruples for all X' and systems (P, L, I) satisfying conditions (1) and (2).

Note that Lemma 2.3 and the minimality condition (3) imply

$$R(P) = R(L), \qquad \mathrm{D}(P) = \mathrm{D}(L) = 1.$$

Set $r = R(P) = R(L)$. Properties (1), (2), (3) are preserved under passage to extensions of X'; we may therefore assume that X' contains all the parameters necessary for the sequel.

Let A be an X-definable strongly minimal set. We claim that $\widehat{A}_X$ is neither projective nor degenerate. Projectivity and degeneracy are inherited upon localization of geometries, so we may assume that $X' = X$. Let $\psi(v_0, v_1)$ be a formula with parameters in X, which is a stratification of L with respect to A of rank less than r, that is,

(4) for any $l \in L$, there exists $a \in A$ such that $\vDash \psi(a, l)$.

(5) For all a in A, $R(\psi(a, \mathfrak{M})) < r$.

We claim that

(6) for all $a \in A$, $R(p_0 I \cap \psi(a, \mathfrak{M})) = 0$.

Indeed, otherwise, putting

$$L_0 = \psi(a, \mathfrak{M}) \cap L, \qquad I_0 = I \cap (P \times L_0),$$

we obtain

$$R(p_0, X \cup \{a\}) \geq r - 1 \geq R(L_0), \qquad R(p_0 I_0) > 0.$$

Obviously, (P, L_0, I_0) satisfies (1) over $X \cup \{a\}$, and (2) follows from Lemma 2.3. This contradicts the minimality of (P, L, I).

Note that, by defining

$$P' = \{p \in P : (\forall a \in A) R(pI \cap \psi(a, \mathfrak{M})) = 0\},$$
$$I' = I \cap (P' \times L),$$

we obtain a system (P', L, I') satisfying conditions (1)–(3). We may therefore assume that $P' = P$.

Let Q be a definable subset of $P \cup L$ of maximum rank among all the Q such that Q is coordinatizable over $A_{X'}$ for some finite $X' \supseteq X$, i.e., $Q \subseteq \operatorname{acl}(A \cup X')$. Since X may be assumed sufficiently large, we can take $X' = X$. Moreover, $Q \neq \varnothing$ because $Q \supseteq (P \cup L) \cap \operatorname{acl}(X)$. Let

$$R(Q \cap L) \leq R(Q \cap P), \qquad L^* = \bigcup \{pI : p \in P \cap Q\}.$$

For any l in L^* there exist p in $P \cap Q$ and a in A such that

$$l \in pI \cap \psi(a, \mathfrak{M}).$$

Since the latter is a set of rank 0, it follows that

$$l \in \operatorname{acl}(\{p, a\} \cup X), \qquad L^* \subseteq \operatorname{acl}(A \cup X),$$

and therefore $R(L^*) \leq R(Q \cap P)$. Choosing $p_0^* \in Q \cap P$ so that

$$R(p_0^*, X) \geq R(L^*),$$

we deduce from Lemma 2.3 that condition (2) holds for (P^*, L^*, I^*), where $P^* = Q \cap P$. Since this system inherits condition (1) from (P, L, I), it follows by minimality that $R(P^*) = R(L^*) = r$. And once again, we may assume that $P^* = P$, $L^* = L$, that is, $P \cup L$ is coordinatizable over A_X.

Using Lemmas 1.2.8, 1.2.9, we obtain: $P \cup L$ is strictly r-coordinatizable over A_X.

We have thus proved:

LEMMA 3.2. *If a pseudoplane is definable in* $\mathfrak{M}$, *then for some finite* $X \subseteq M$ *an incidence system* (P, L, I), *such that* P *and* L *are strictly* r-*coordinatizable over* A_X *and conditions* (1)–(3) *are satisfied, is* X-*definable in* $\mathfrak{M}$.

We now continue the proof of Proposition 3.1. To simplify the notation, let us assume that $X = \varnothing$. Then

$$\operatorname{acl}(p_0) = \operatorname{acl}(a_1, \ldots, a_r), \quad \operatorname{acl}(l_0) = \operatorname{acl}(b_1, \ldots, b_r),$$
$$a_1, \ldots, a_r, b_1, \ldots, b_r \in A.$$

Since

$$R(p_0, \varnothing) = r, \qquad R(l_0, \varnothing) = r,$$

it follows that

$$\dim A[a_1, \ldots, a_r] = r = \dim A[b_1, \ldots, b_r].$$

It follows from condition (1) that

$$R(p_0, \{l_0\}) < r,$$

and therefore $a_1, \ldots, a_r$ are dependent over $b_1, \ldots, b_r$, i.e.,

$$\dim A[a_1, \ldots, a_r, b_1, \ldots, b_r] < 2r.$$

We shall show that for a degenerate or projective geometry $\widehat{A}$ the condition

$$\dim A[\{a_1, \ldots, a_r\} \cup Y] < \dim A[a_1, \ldots, a_r] + \dim A[Y]$$

implies the existence of $a \in A - \operatorname{acl}\{\varnothing\}$ such that

$$a \in A[a_1, \ldots, a_r] \cap A[Y].$$

We prove this by induction on r, assuming $a_1, \ldots, a_r$ to be independent. If $r = 1$ we may clearly take $a = a_1$. If $r > 1$,

$$a_1 \in \operatorname{acl}(\{a_2, \ldots, a_r\} \cup Y).$$

We use condition (1.6). There exists $a \in A$ such that $a \in \operatorname{acl}(\{a_3, \ldots, a_r\} \cup Y)$, $a_1 \in \operatorname{acl}(a_2, a)$, that is, $a \in \operatorname{acl}(a_1, a_2)$. Hence it follows that

$$\dim A[\{a, a_3, \ldots, a_r\} \cup Y] < \dim A[a, a_3, \ldots, a_r] + \dim A[Y].$$

Therefore, by the induction hypothesis, there exists $d \in A - \operatorname{acl}(\varnothing)$ such that

$$d \in \operatorname{acl}(a, a_3, \ldots, a_r) \cap \operatorname{acl}(Y).$$

Since $a \in \operatorname{acl}(a_1, a_2)$, this proves our assertion.

Thus, if $\widehat{A}$ is a degenerate or projective geometry, there exists

$$a \in (A - \operatorname{acl}(\varnothing)) \cap A[a_1, \ldots, a_r] \cap A[b_1, \ldots, b_r],$$

i.e.,

$$a \in \operatorname{acl}(p_0) \cap \operatorname{acl}(l_0).$$

Hence $R(l_0, \{a\}) < r$.

This inequality readily yields a $\varnothing$-definable $\varphi(v_0, v_1)$ such that $\varphi(a, l_0)$ is true and $R(\varphi(a, \mathfrak{M})) < r$. Since $R(l_0, \varnothing) = r$, the formula φ may be adjusted in such a way as to obtain (4) and (5). Hence (6) is also true for the selected φ. But the set $p_0 I \cap \varphi(a, \mathfrak{M})$ contains l_0, and therefore $l_0 \in \mathrm{acl}(p_0, a) = \mathrm{acl}(p_0)$. This contradicts (2). □

PROPOSITION 3.3. *If the geometry $\widehat{A}$ of a strongly minimal countably categorical structure $\mathfrak{A}$ is neither degenerate nor locally projective, then for some finite $X \subseteq A$ there is a pseudoplane (P, L, I), X-definable in $\mathfrak{A}$, such that*

(1) $\mathrm{RD}(P) = \mathrm{RD}(L) = \langle 2, 1\rangle$;

(2) $R(pI) = 1$, $R(Il) = 1$, *for all* $p \in P$, $l \in L$;

(3) P *and* L *are strictly 2-coordinatizable over* A_X;

(4) *if* $\mathfrak{A}$ *is countably categorical, then* P *and* L *are atoms over* X.

PROOF. It follows from Proposition 2.5 that a pseudoplane with property (1) is definable in $\mathfrak{A}$. Condition (2) follows from (1) by rank evaluation. Condition (3) follows from Lemma 3.2 because $r = 2$.

If $\mathfrak{A}$ is countably categorical then, taking X-atoms P', L' such that $P' \subseteq P$, $L' \subseteq L$, $R\mathrm{D}(P') = R\mathrm{D}(L') = \langle 2, 1\rangle$, $I' = I \cap (P' \times L')$, we again obtain a pseudoplane (P', L', I') satisfying all of conditions (1)–(4). □

PROPOSITION 3.4. *Let $\mathfrak{A}$ and $\mathfrak{A}'$ be strongly minimal structures, definable in an uncountably categorical structure $\mathfrak{M}$. Then for some finite X the geometries $\widehat{A}_X$ and $\widehat{A}'_X$ are isomorphic.*

PROOF. According to 1.2.3, there exists an X-definable stratification φ of the set A' with respect to A. Let $\mathfrak{A}$, $\mathfrak{A}'$ be definable over X; to simplify matters, let us assume that $X = \varnothing$.

For any $a' \in A' - \mathrm{acl}(\varnothing)$ there exists $a \in A - \mathrm{acl}(\varnothing)$ such that $\varphi(a, a')$ and $R(a', \{a\}) = 0$. Then by the reciprocity law $R(a, \{a'\}) = 0$, i.e., $\mathrm{acl}(a) = \mathrm{acl}(a')$. Hence it follows that φ realizes an isomorphism of the geometries $\widehat{A}$ and $\widehat{A}'$. □

COROLLARY. *$\widehat{A}$ is degenerate if and only if $\widehat{A}'$, is degenerate.*

Indeed, it is easy to see that $\widehat{A}$ is degenerate if and only if $\widehat{A}_X$ is degenerate.

THEOREM 3.5 (Trichotomy Theorem). *For an uncountably categorical structure $\mathfrak{M}$, one and only one of the following three conditions is fulfilled*:

(1) *There is a pseudoplane definable in* $\mathfrak{M}$.

(2) *The geometry of any strongly minimal structure definable in $\mathfrak{M}$ is locally projective.*

(3) *The geometry of any strongly minimal structure definable in $\mathfrak{M}$ is degenerate.*

Conditions (2) *and* (3) *are equivalent, respectively, to the conditions*:

(2′) *The geometry of some strongly minimal structure definable in* $\mathfrak{M}$ *is locally projective.*

(3′) *The geometry of some strongly minimal structure definable in* $\mathfrak{M}$ *is degenerate.*

PROOF. Suppose that $\mathfrak{M}$ does not satisfy condition (1). Then by Proposition 2.5 the geometry of any strongly minimal set A definable in $\mathfrak{M}$ is locally projective or degenerate. By the corollary to Proposition 3.4, if $\widehat{A}$ for some A is degenerate, then $\widehat{A}$ is degenerate for all A. Thus either condition (2) or (3) holds, and these conditions are equivalent to (2′) or (3′), respectively. □

Structures satisfying condition (1) will henceforth be called *field-like*, those satisfying (2) *module-like* and those satisfying (3) *of disintegrated type.*

§4. Definable groups in different types of uncountably categorical structures

As we shall show later (Chapter 5), the global structure of a structure $\mathfrak{M}$ is largely characterized by the groups definable in $\mathfrak{M}$. In this connection, the theorems about groups in this section are of added interest.

THEOREM 4.1. *Let* G *be a group, definable in an uncountably categorical structure* $\mathfrak{M}$ *which is module-like or of discrete type. Then,*

(a) G *is almost abelian*;

(b) *if* G *is infinite and has no infinite definable proper subgroups, then* G *is strongly minimal.*

PROOF. We begin with part (b). We shall show that for any strongly minimal subset Q of G the difference $G - A$ is finite. Define

$$H = \{h \in G : Q \cdot h \sqsupset\sqsubset Q\},$$
$$H' = \{h \in G : h \cdot Q \sqsupset\sqsubset Q\}.$$

Then, by 1.4.5, 1.4.6, H and H' are definable subgroups of G and $gH \sqsubset Q$, $H'g' \sqsubset Q$ for some $g, g' \in G$. If either H or H' is infinite, it coincides with G, and so $G - Q$ is finite. We may therefore assume that both H and H' are finite. By definition, $Q \sqsupset\sqsubset H'QH$, so we may assume that $Q = H'QH$.

Define

$$P = \{g \cdot H' : g \in G\}, \qquad L = \{H \cdot g : g \in G\},$$
$$I = \{\langle g' \cdot H', H \cdot g\rangle : g' \in Q \cdot g\}.$$

It is easy to verify directly that (P, L, I) is a pseudoplane. By the Trichotomy Theorem this contradicts the assumption, proving part (b).

The proof of part (a) is divided into several lemmas.

LEMMA 4.2. *Let* $J \subseteq U \times G$ *be a two-place relation and let* U *be a set, both definable in* G, *such that* uJ *is a strongly minimal subgroup of* G *for any* u *in* U, *and for different* u_1, u_2 *in* U, $u_1 J \neq u_2 J$. *Then* U *is finite.*

Proof. We shall show that if U is infinite, then there is a pseudoplane definable in G.

We may assume that U is a strongly minimal set. We define

$$H = \{g \in G : R(Jg) > 0\},$$
$$P = \{H \cdot g : g \in G\}, \qquad L = \{g \cdot uJ : g \in G,\ u \in U\},$$
$$I = \{\langle g \cdot H,\ g \cdot uJ\rangle : g \in G,\ u \in U\}.$$

The set H is finite. Indeed, for any $h_1, \dots, h_k \in H$ we have $R(Jh_1 \cap \dots \cap Jh_k) > 0$, and so we can find distinct $u_1, u_2 \in U$ such that

$$u_1 J \cap u_2 J \supseteq \{h_1, \dots, h_k\}.$$

If H were infinite, it would follow from the Compactness Theorem that the subgroup $u_1 J \cap u_2 J$ is infinite for distinct u_1, u_2 in U, and since $u_1 J$ and $u_2 J$ are strongly minimal, this would imply $u_1 J = u_2 J$, which is impossible.

H is a subgroup of G. Indeed, if $h_1, h_2 \in H$ then $J(h_1 \cdot h_2^{-1}) \supseteq Jh_1 \cap Jh_2$, since uJh_1 and uJh_2 imply $h_1 \cdot h_2^{-1} \in uJ$, whence $u \in J(h_1 \cdot h_2^{-1})$.

It is now easy to verify directly that (P, L, I) is a pseudoplane. □

Lemma 4.3. *G has a definable normal nilpotent subgroup of finite index.*

Proof. We may assume that G is infinite and identical with its own connected component, i.e., that G is connected. Then the group $G \times G$ is also connected. Let H be a strongly minimal subgroup of G—such a subgroup exists by the already proved part (b) of the theorem. Denote

$$gI\langle h, h'\rangle \Leftrightarrow h \in H \ \&\ h' = g^{-1} \cdot h \cdot g.$$

Clearly, for any $g \in G$ the set gI is a subgroup of $G \times G$ isomorphic to H, that is, it is strongly minimal. By Lemma 4.2, G/E is finite, where $g_1 E g_2$ stands for $g_1 I = g_2 I$ for all $g_1, g_2 \in G$. Consequently, the centralizer of H in G has finite index in G, and since G is connected this centralizer must be all of G. We have thus shown that a connected group G has an infinite center (containing H), whence we deduce by induction on $R(G)$ that G is nilpotent. □

Lemma 4.4. *G has a definable abelian normal subgroup of finite index.*

Proof. We shall prove that if G is connected, then it is abelian.

As shown in the previous lemma, G is nilpotent. Assuming that the nilpotency class of G is at most 2 (that is to say, $G/C(G)$ is abelian), we shall show that G is abelian.

Thus, let $\overline{G} = G/C(G)$ be an abelian group. Since G is connected, $\overline{G}$ is also connected and is therefore either infinite or trivial. We claim that the first possibility implies a contradiction. Let $\overline{H}$ be a strongly minimal subgroup of $\overline{G}$ (existing by part (b)). Set

$$gI\langle \bar{h}, \bar{h}'\rangle \Leftrightarrow \bar{h}, \bar{h}' \in H \ \&\ g \in G \ \&\ \bar{h}' = \bar{h} \cdot g \cdot \bar{h}^{-1} \cdot g^{-1}.$$

It is easy to see that for any $g \in G$ the set gI is a subgroup of the group $\overline{G} \times C(G)$, isomorphic to the strongly minimal group $\overline{H}$. By Lemma 4.2, G/E is finite, where $g_1 E g_2$ stands for $g_1 I = g_2 I$, i.e., $g_1 \cdot \bar{h} \cdot g_1^{-1} = g_2 \cdot \bar{h} \cdot g_2^{-1}$ for all $\bar{h} \in \overline{H}$. In other words, the centralizer of the subgroup $H = \{h \in G : h \cdot C(G) \in \overline{H}\}$ in G has finite index, and thus coincides with G. Therefore $H = C(G)$, $\overline{H} = \bar{1}$. Contradiction. This proves the lemma and completes the proof of Theorem 4.1. □

THEOREM 4.5. *There are no infinite groups definable in uncountably categorical structures of discrete type.*

PROOF. Let $\mathfrak{M}$ be a structure of discrete type, G a group definable in $\mathfrak{M}$. By Theorem 4.1(b), if G is infinite, there exists a strongly minimal subgroup of G. To prove the theorem, therefore, we need only show that no strongly minimal group G is definable in $\mathfrak{M}$.

Supposing, then, that G is a strongly minimal group, we may assume that it is $\varnothing$-definable in $\mathfrak{M}$. Let g, h be algebraically independent elements of G and $f = g \cdot h$. Obviously, $f \in \operatorname{acl}(g, h)$, but $f \notin \operatorname{acl}(g)$, $f \notin \operatorname{acl}(h)$. Since $\mathfrak{M}$ is of disintegrated type,

$$G[g, h] = G[g] \cup G[h],$$

contrary to the previous remark. □

CHAPTER III

Classification of Infinite Locally Finite Homogeneous Pregeometries

In this chapter we present the proof of one of the main theorems of this dissertation. This classification theorem describes all the infinite locally finite homogeneous geometries, up to isomorphism.

As shown in Chapter 2, §1, one can associate to every pregeometry A a geometry $\widehat{A}$. It is obvious that homogeneous locally finite pregeometries A are characterized up to isomorphism by the isomorphism type of the homogeneous locally finite geometry $\widehat{A}$ and two numbers: the number of elements in $\mathrm{acl}(\emptyset)$ and the (total) number of elements in $\mathrm{acl}(x)$ for all $x \in A - \mathrm{acl}(\emptyset)$.

Thus, the classification theorem will also describe the infinite locally finite homogeneous pregeometries and thus, by Proposition 2.1.1, the pregeometries of strongly minimal sets in totally categorical theories.

THEOREM 0.1 (Classification Theorem). *An arbitrary infinite locally finite homogeneous geometry is isomorphic to a degenerate, affine or projective geometry over a finite field.*

COROLLARY 0.2. *Totally categorical structures are either module-like or of discrete type.*

§1. Quasidesigns

In this section we study incidence systems $(P, L, I) = \mathscr{P}$ together with a strongly minimal set A, which are $\emptyset$-definable in a totally categorical structure $\mathfrak{M}$ and satisfy the following conditions:

(1) for all $l \in L$, $R(Il) = 1$;

(2) for all $l_1, l_2 \in L$, if $l_1 \neq l_2$, then $R(Il_1 \cap Il_2) = 0$;

(3) $RD(P) = RD(L) = \langle 2, 1 \rangle$;

(4) P and L are atoms;

(5) P and L are strictly 2-coordinatizable over A.

Such a system will be called a *quasidesign over* A. This term is motivated by the notion of "design", widely used in combinatorial analysis: a finite incidence system (P, L, I) in which $\|Il\| = \kappa$, $\|p_1 I \cap p_2 I\| = \mu$, $\|P\| =$

$\pi \geq \kappa + 2$ for some nonzero natural numbers κ, μ, π and any $l \in L$, $p_1, p_2 \in P$, $p_1 \neq p_2$ [46].

It is clear that the pseudoplane (P, L, I) of Proposition 2.3.3 is a quasidesign over A if X is added to the signature. Not every quasidesign is a pseudoplane, but the following result is true.

LEMMA 1.1. *Conditions* (1), (2), (3), *and* (4) *imply the following conditions*:

(6) *for all* $p \in P$, $R(pI) = 1$;

(7) *for all* $p_1, p_2 \in P$, *if* $p_2 \notin \operatorname{acl}(p_1)$, *then* $R(p_1 I \cap p_2 I) = 0$.

PROOF. (6) Let $p \in P$, $l \in L$, and pIl. Since $R(p, \{l\}) + R(l, \varnothing) = R(l, \{p\}) + R(p, \varnothing)$ this implies, in view of (3) and (4), that

$$R(p, \{l\}) = R(l, \{p\}).$$

In view of (1), $R(p, \{l\}) = 1$ for a suitable $p \in Il$, and this value is maximal. In this case

$$R(l, \{p\}) = R(pI) = 1,$$

and by (4) this gives (6).

(7) Suppose that $p_2 \notin \operatorname{acl}(p_1)$ but $R(p_1 I \cap p_2 I) = 1$. Let $l_1, l_2 \in p_1 I \cap p_2 I$, where also

$$R(l_1, \{p_1\}) = R(l_1, \{p_1, p_2\}) = 1, \qquad R(l_2, \{p_1\}) = R(l_2, \{p_1, p_2, l_1\}) = 1.$$

Using Lascar's Reciprocity Principle, we obtain

$$R(p_2, \{l_1, l_2, p_1\}) \geq 1.$$

This contradicts condition (2), since $p_2 \in Il_1 \cap Il_2$ and $l_1 \neq l_2$. □

Our main tool for investigating quasidesigns will be the Polynomial Theorem 1.5.5, which will be applied throughout this chapter under the assumption of strict coordinatizability, i.e., on the assumption that $n_A = 0$.

We proceed to the study of a quasidesign $\mathscr{P}$.

Let X be an arbitrary finite closed set in $\mathfrak{M}$. Throughout we shall denote

$$x = \|A[X]\|.$$

The finite incidence system

$$(P[X], L[X], I[X])$$

will be denoted by $\mathscr{P}[X]$. Our examination of combinatorial relations in $\mathscr{P}[X]$ and the terminology will adhere to the classical monograph [46, Chapters 1, 2 and 7].

By Theorem 1.5.5, there exist polynomials π and λ, independent of X, such that

$$\|P[X]\| = \pi(x), \qquad \|L[X]\| = \lambda(x).$$

In addition, the number of points of $P[X]$ on a line $l \in L[X]$ is a polynomial independent of l, in view of condition (4) and part (d) of Theorem 1.5.5. Denote this polynomial by $\kappa(x)$. Similarly, the polynomial $\rho(x)$ computes the number of lines in $L[X]$ that pass through a given point of $P[X]$ (this shows that $\mathscr{P}[X]$ is a tactical configuration [46]).

It is easy to see that the following relation holds:

(8) $\lambda \cdot \kappa = \pi \cdot \rho$.

According to Theorem 1.5.5(c) and the definition of a quasidesign,

(9) $\deg \lambda = \deg \pi = 2$,

(10) $\deg \kappa = \deg \rho = 1$.

Let Y be a finite set and U a Y-definable set. Countable categoricity implies

$$U = \bigcup \{U_j : j \in J(U)\},$$

where $U_1, \dots, U_k$ are distinct atoms over Y and $J(U) = \{1, \dots, k\}$ is an index set. We shall refer to this partition as a partition into elementary equivalence classes over Y.

Let

$$S = \{\langle p_1, p_2 \rangle : p_1, p_2 \in P, \ p_1 \neq p_2\},$$

and let

$$S = \bigcup \{S_j : j \in J(S)\}$$

be a partition of S into elementary equivalence classes over $\varnothing$. This partition induces a partition

(11) $S[X] = \{S_j[X] : j \in J(S)\}$.

By Theorem 1.5.5, for any $p_0 \in P[X]$ the number of points $p \in P[X]$ such that $\langle p, p_0 \rangle \in S_j$ for fixed $j \in J(S)$ is expressed by a polynomial which does not depend on p_0. Denote this polynomial by $\sigma_j(x)$. It is easy to see that

(12) $\sum_{j \in J(S)} \sigma_j = \pi - 1$.

Since P cannot be partitioned into two definable subsets of rank 2, only one of the σ_j's can be of degree 2. To fix ideas, let us assume that

(13) $\deg \sigma_1 = 2$ and $\deg \sigma_j \leq 1$ for $j \neq 1$, $j \in J(S)$.

These relations mean, in particular, that for fixed $p_0 \in P$

$$\begin{gathered} R(\{p \in P : \langle p, p_0 \rangle \in S_1\}) = 2, \\ R(\{p \in P : \langle p, p_0 \rangle \in S_j\}) \leq 1, \quad \text{for } j \neq 1, \ j \in J(S). \end{gathered} \tag{13$'$}$$

Let μ_j denote the number of lines in $L[X]$ passing through a pair of points p_1, p_0 such that $\langle p_1, p_0 \rangle \in S_j[X]$. It follows from (10) that $\deg \mu_j \leq 1$, because $\mu_j(x) \leq \rho(x)$. It follows from (7) that

(14) if $\deg \mu_j \geq 1$, then $\deg \mu_j = 1$ and $\deg \sigma_j = 0$.

Computing the number of elements of the set

$$\{\langle p, l \rangle \in I[X] : l \in p_0 I, \ p \neq p_0\}$$

in two different ways for fixed $p_0 \in P[X]$, we obtain

(15) $\sum_{j \in J(S)} \mu_j \sigma_j = (\kappa - 1) \cdot \rho$.

It follows from (13), (14), and (15) that

(15′) $\deg \mu_1 \cdot \sigma_1 = 2$, $\deg(\kappa \cdot \rho - \mu_1 \cdot \sigma_1) < 2$.

Let Y be finite, $Y \subseteq X$. Define a Y-decomposition of the quasidesign $\mathscr{P}$ to be the partitions

(16) $P = \bigcup\{P_i : i \in J(P)\}$, $L = \bigcup\{L_j : j \in J(L)\}$

into elementary equivalence classes over Y. The partitions (16) induce corresponding partitions of $P[X]$ and $L[X]$.

Denote

$$\|P_i[x]\| = \pi_i(x), \qquad \|L_j[X]\| = \lambda_j(x).$$

As before, we observe that in each of the sets of polynomials π_i and λ_j there is exactly one polynomial of degree 2. To fix ideas, let us assume that

(17a) $\deg \pi_1 = 2$, $\deg \pi_j \leq 1$ for $j \neq 1$, $j \in J(P)$;

(17b) $\deg \lambda_1 = 2$, $\deg \lambda_j \leq 1$ for $j \neq 1$, $j \in J(L)$.

The following relations are also obvious:

(18a) $\sum_{i \in J(P)} \pi_i = \pi$,

(18b) $\sum_{j \in J(L)} \lambda_j = \lambda$.

It follows from the definition of a Y-decomposition and from Theorem 1.5.5 that the number of points in $P_i[X]$ lying on a line l in $L_j[X]$ depends on i, j, and x, but not on l. Denote this number by $\pi_{ij}(x)$. Similarly, denote the number of lines in $L_j[X]$ passing through a fixed point in $P_i[X]$ by $\lambda_{ji}(x)$.

It is easy to see that

(19a) $\sum_{i \in J(P)} \pi_{ij} = \kappa$ for all $j \in J(L)$,

(19b) $\sum_{j \in J(L)} \lambda_{ji} = \rho$ for all $i \in J(P)$.

Finally, we observe that for any $i \in J(P)$ and $j \in J(L)$, $P_i[X]$ and $L_j[X]$ form a tactical configuration, that is, a Y-decomposition is a tactical decomposition. Hence we have the following relation, analogous to (8):

(20) $\lambda_j \cdot \pi_{ij} = \pi_i \cdot \lambda_{ji}$.

For $i \in J(P)$, $p_0 \in P$, let us denote

$$G_{ip_0} = \{\langle p, l\rangle \in I : l \in p_0 I,\ p \in P_i,\ p \neq p_0\}$$

and $\gamma_{ik}(x) = \|G_{ip_0}[X]\|$ for $p_0 \in P_k[X]$.

Obviously,

$$\gamma_{ij} = \sum_{j \in J(L)} \gamma_{ik}^j,$$

where

$$\gamma_{ik}^j = \|\{\langle p, l\rangle \in G_{ip_0}[X] : l \in L_j\}\|, \qquad p_0 \in P_k[X].$$

In turn,

$$\gamma_{ik}^j = \lambda_{jk} \cdot (\pi_{ij} - \delta_{ik}),$$

where δ_{ik} is the Kronecker symbol: $\delta_{ik} = 0$ for $i \neq k$ and $\delta_{ii} = 1$.

From this and (19b) we conclude that

(21) $\gamma_{ik} = (\sum_{j \in J(L)} \pi_{ij} \cdot \lambda_{jk}) - \rho\delta_{ik}$.

In the sequel, given polynomials f and g, we write $f \sqsupset g$ if $\deg(f-g) < \deg g$, where $f \sqsubset g$ will mean that either $f \sqsupset g$ or the leading coefficient of $f-g$ is negative.

Proposition 1.2. *Suppose that in the Y-decomposition* (16) P_2 *is a strongly minimal set and the following condition holds:*
(22) *for any line* $l \in L$, $Il \cap P_2$ *is finite.*
Then
(23) $\rho \sqsubset (\mu_1 - \nu) \cdot \pi_2$,
where ν *is the number of lines incident with a pair of points of* P_2 *independent over* Y.

Remark. The number ν is independent of the choice of the pair of points, since P_2 is strongly minimal, In addition, $\deg \nu = 0$ by Lemma 1.1, and $\deg \mu_1 = 0$ by (13) and (14).

Proof (of Proposition 1.2). Condition (22) means that for all $j \in J(L)$
(22′) $\deg \pi_{2j} = 0$.
Computing the degrees of the left- and right-hand sides of (20) for $i = 2$ and $j \in J(L)$, we deduce from (18) and (17b) that
(24) $\deg \lambda_{12} = 1$, $\deg \lambda_{j2} = 0$ for $j \neq 1$, $j \in J(L)$,
and together with (19b) and (22) this implies
(25) $\lambda_{12} \sqsupset \rho$,
(26) $\sum_{j \in J(L)} \pi_{2j} \cdot \lambda_{j2} \sqsupset \pi_{21} \cdot \rho$,
(26′) $\gamma_{22} \sqsupset \pi_{21} \cdot \rho - \rho$.
We now show that
(27) $\pi_{21} \cdot \rho \sqsupset \mu_1 \cdot \pi_2$.
To that end, we note that by (20) and (25)

$$\lambda_1 \cdot \pi_{21} \sqsupset \pi_2 \cdot \rho,$$

whence

$$\lambda_1 \cdot \pi_{21} \cdot \kappa \sqsupset \pi_2 \cdot \rho \cdot \kappa,$$

but, by (15′),

$$\rho\kappa \sqsupset \mu_1 \cdot \sigma_1,$$

and therefore

$$\lambda_1 \cdot \pi_{21} \cdot \kappa \sqsupset \mu_1 \cdot \sigma_1 \cdot \pi_2.$$

On the other hand, $\lambda_1 \sqsupset \lambda$, $\sigma_1 \sqsupset \pi$ (see (12), (13), (17b), (18b)), whence it follows that

$$\pi_{21} \cdot \lambda \cdot \kappa \sqsupset \mu_1 \cdot \pi_2 \cdot \pi,$$

or, in view of (8),

$$\pi_{21} \cdot \pi \cdot \rho \sqsupset \mu_1 \cdot \pi_2 \cdot \pi,$$

which implies (27).
We now refine (26′):
(28) $\gamma_{22} \sqsupset \mu_1 \cdot \pi_2 - \rho$.

Let us now calculate γ_{22} in another way. To that end, fixing some point $p_0 \in P_2[X]$, let us consider the set V_{p_0} of points of P_2 which are independent of p_0 over Y. Let $v(x) = \|V_{p_0}[X]\|$. Since P_2 is strongly minimal, it follows that $P_2 - V_{p_0}$ is finite, and therefore $v \sqsupset\!\sqsubset \pi_2$. Define

$$G'_{2p_0} = \{\langle p, l\rangle \in I:\ p_0 I l,\ p \in V_{p_0}\}, \qquad \gamma'_{22}(x) = \|G'_{2p_0}[X]\|.$$

Clearly, $\gamma'_{22}(x) = v(x) \cdot \nu$. Since $G_{2p_0} \supseteq G'_{2p_0}$, it follows that

$$\gamma_{22} \sqsupset v \cdot \nu \sqsupset\!\sqsubset \pi_2 \cdot \nu.$$

Comparing this with (28), we obtain (23). □

PROPOSITION 1.3. *For any quasidesign*, $\kappa \sqsupset \rho$ *and* $\pi \sqsupset \lambda$.

PROOF. Fixing $l_0 \in L$, we can find a finite Y such that $l_0 \in Y$ and a strongly minimal subset of Il_0 is definable relative to Y. Suppose that in the Y-decomposition (16) the set P_2 is strongly minimal and $P_2 \subseteq Il_0$.

Obviously,

(29) $\pi_2 \sqsubset \kappa$.

Let $l_0 \in L_n$. By condition (2), l_0 is the only line that intersects P_2 in an infinite set, and, since L_n is an elementary equivalence class, it follows that $L_n = \{l_0\}$. That is to say,

(30) $\lambda_n = 1$, $\pi_{2n} = \pi_2$, $\lambda_{n2} = 1$;

(31) $\deg \pi_{2j} = 0$ if $j \neq n$, $j \in J(L)$.

Calculating the degrees of the left- and right-hand sides of (20) for $i = 2$, with due allowance for (17b), (19b), (30) and (31), we again obtain (24) and (25). Formula (21) together with (30), (31), and (24) now gives

$$\gamma_{22} \sqsupset\!\sqsubset \pi_{21} \cdot \rho + \pi_2 - \rho.$$

As shown previously, formula (25) implies (27), and so

(32) $\gamma_{22} \sqsupset\!\sqsubset (\mu_1 + 1) \cdot \pi_2 - \rho$.

On the other hand, independently of condition (22) it is true that $\gamma_{22} \sqsupset \nu \cdot \pi_2$, where ν is the number of lines passing through a pair of points $\{p_0, p_1\} \subseteq P_2$ such that $p_1 \in V_{p_0}$, $V_{p_0} = P_2 - \operatorname{acl}(\{p_0\} \cup Y)$.

We now show that $\mu_1 = \nu$. Consider a point $p \in V_{p_0}$. Recall that by (13) $\mu_1 = \nu$ if $\langle p, p_0\rangle \in S_1$, i.e., $R(p, \{p_0\}) = 2$.

Since $p \in V_{p_0}$, V_{p_0} is infinite and V_{p_0} is an atom over $Y \cup \{p_0\}$, it follows that

$$R(p, Y \cup \{p_0\}) \geq 1,$$

and moreover

$$R(p, \{l_0, p_0\}) \geq 1.$$

Taking into consideration that $p \in Il_0$ and $R(Il_0) = 1$, we obtain

$$R(p, \{l_0, p_0\}) = 1.$$

If we now assume that $R(p, \{p_0\}) = 1$, we see that p and l_0 are independent over $\{p_0\}$ in Lascar's sense. But this is impossible, because

$$R(l_0, \{p, p_0\}) = 0$$

in view of condition (7), and

$$R(l_0, \{p_0\}) = 1.$$

We have thus shown that $\mu_1 = \nu$ and we finally obtain

(33) $\gamma_{22} \sqsupset \mu_1 \cdot \pi_2$.

Comparing this with (32), we have $\pi_2 \sqsupset \rho$.

Taking (29) into account we get $\kappa \sqsupset \rho$, and by (8) $\pi \sqsupset \lambda$. □

PROPOSITION 1.4. *An arbitrary strongly minimal Y-definable subset $Q \subseteq P$ satisfies the condition*

(34) *$Q - Il\,a$ is finite for some $l \in L$,*

or

(35) *if $p_0 \in Q - \mathrm{acl}(Y)$, then $Q^{(1)}[p_0]$ is infinite.*

PROOF. We note that $P_2 = Q - \mathrm{acl}(Y)$ is an atom over Y. If (22) is not true for P_2, then $Il \cap Q$ is infinite for some $l \in L$, and since Q is strongly minimal, this gives (34).

If (22) is true for P_2, then in view of (23) the number of lines of L incident with points $p_1, p_0 \in Q - \mathrm{acl}(Y)$, where $p_1 \notin \mathrm{acl}(Y \cup \{p_0\})$, is not equal to μ_1. By definition, this means that $\langle p_1, p_0\rangle \notin S_1$. Then, by (13), we have $R(p_1, \{p_0\}) = 1$, $Q - \mathrm{acl}(Y \cup \{p_0\}) \subseteq Q^{(1)}[p_0]$. □

§2. Semiprojectivity

We shall say that a geometry A is *semiprojective* if for any $x \in A$, $Z \subseteq A$, $y \in A[\{x\} \cup Z]$ there exist $z_1, z_2 \in A[Z]$ such that $y \in A[x, z_1, z_2]$.

THEOREM 2.1. *For any locally finite infinite homogeneous pregeometry A there exists a finite subset $X \subseteq A$ such that $\widehat{A}_X$ is a semiprojective geometry.*

PROOF. It is easy to see that if the geometry $\widehat{A}$ is degenerate or locally projective, then $\widehat{A}$ is semiprojective. Therefore, by the Trichotomy Theorem, we may assume that some pseudoplane (P, L, I) is X-definable in $\mathfrak{A}$; in fact, by Proposition 2.3.3, we may even assume that this pseudoplane is a quasidesign over A_X. Fixing X, we may assume from now on, for simplicity of notation, that $X = \varnothing$.

Let $x_0 \in A - A[\varnothing]$, $Z \subseteq A$, $y_0 \in A[\{x_0\} \cup Z] - A[x_0]$, $p_0 \in P$, $A[p_0] = A[x_0, y_0]$, $\dim A[Z] \geq 2$. Consider the following subset S of P:

$$S = \{p \in P\colon \mathrm{tp}(p, A[Z]) = \mathrm{tp}(p_0, A[Z])\}.$$

By construction, S is an atom over $A[Z]$, and since

$$R(p_0, A[Z]) = R(\langle x_0, y_0\rangle, A[Z]) = 1,$$

it follows that $R(S) = 1$. Our problem is to find $z_1, z_2 \in A[Z]$ such that $S^{(1)}[z_1, z_2] \neq \varnothing$. Hence it follows at once, since S is an $A[Z]$-atom, that $1 = R(p_0, \{z_1, z_2\}) = R(\langle x_0, y_0\rangle, \{z_1, z_2\})$, $R(y_0, \{x_0, z_1, z_2\}) < 1$, that is, $y_0 \in A[x_0, z_1, z_2]$, which is the condition for semiprojectivity. □

We shall need the following auxiliary proposition.

LEMMA 2.2. *Let $Q \subseteq S$, where Q is a strongly minimal atom over some finite Z', $Z' \supseteq A[Z]$ and $Q^{(1)}[p]$ is infinite for some $p \in Q$. Then either there exists $z \in A[Z]$ such that $S^{(1)}[z] \neq \varnothing$ or there exists $q \in P[Z]$ such that $S^{(1)}[q] \neq \varnothing$.*

PROOF. Since Q is an atom, it follows that $Q^{(1)}[p]$ is infinite for all $p \in Q$. We may assume that $p_0 \in Q$. We shall also assume from the start that $S^{(1)}[z] = \varnothing$ for all $z \in A[Z]$.

Thanks to countable categoricity, there exists m such that for all $q \in P$, if $\|S^{(1)}[q]\| > m$, then $S^{(1)}[q]$ is infinite.

Inductively, let us choose different $p_0, p_1, \dots, p_m \in Q$ so that for all $i < m$,

$$R(p_{i+1}, A[Z] \cup \{p_0, \dots, p_i\}) = R(p_i, A[Z]) = 1.$$

Then by induction, using the Reciprocity Principle, we see that for every $z \in A[Z]$

$$R(z, \{p_0, \dots, p_m\}) = R(z, \{p_0\}).$$

But $R(z, \{p_0\}) = R(z, \varnothing)$, for otherwise $p_0 \in S^{(1)}[z]$. Thus,

$$R(z, \{p_0, \dots, p_m\}) = R(z, \varnothing),$$

that is,

$$A[p_0, \dots, p_m] \cap A[Z] = A[\varnothing].$$

On the other hand, $Q^{(1)}[p_0] \cap \cdots \cap Q^{(1)}[p_m]$ is infinite; let p be a point of this intersection, $A[p] \nsubseteq A[p_0, \dots, p_m]$. Then for fixed $z_1 \in A[Z] - A[\varnothing]$ we can find $q \in P$ such that $z_1 \in A[q]$ and $\operatorname{tp}(q, \{p_0, \dots, p_m\}) = \operatorname{tp}(p, \{p_0, \dots, p_m\})$. In particular, $R(p_i, \{q\}) \leq 1$ for all $i \leq m$, since $R(q, \{p_i\}) = R(p, \{p_i\}) = 1 < R(q, \varnothing)$. Hence $\|S^{(1)}[q]\| > m$, and therefore $S^{(1)}[q]$ is infinite.

Let $p \in S^{(1)}[q]$, $A[p] = A[x, y]$, $x, y \in A$, $x \notin \operatorname{acl}(Z)$. Since

$$R(p, \{q, z_1\}) \leq 1 < R(p, \{z_1\}),$$

it follows that

$$R(q, \{p, z_1\}) < R(q, \{z_1\}) = 1,$$

that is, $q \in \operatorname{acl}(p, z_1)$. But $p \in S$, and therefore $p \in \operatorname{acl}(\{x\} \cup Z)$, and so $q \in \operatorname{acl}(\{x\} \cup Z)$, that is, $R(q, \{x\} \cup Z) = 0$. But $R(x, \{q\} \cup Z) = 1 = R(x, Z)$, so that by the Reciprocity Principle $R(q, z) = 0$. □

We now continue the proof of Theorem 2.1.

Let us choose a strongly minimal subset Q of S which is definable over some finite Z', $Z' \supseteq A[Z]$; we may assume that this Q is a Z'-atom.

By Proposition 1.4, Q satisfies one of two conditions. If condition (1.34) holds, then $Il_0 \cap Q$, and consequently also $Il_0 \cap S$ and $S^{(1)}[l_0]$, are infinite for some $l_0 \in L$. It follows from condition (1.2) and the fact that $R(S) = 1$ that the set

$$\{l \in L\colon R(Il \cap S) > 0\}$$

is finite. Therefore, $l_0 \in \operatorname{acl}(Z)$. But $\operatorname{acl}(l_0) = \operatorname{acl}(z_1, z_2)$ for some $z_1, z_2 \in A$. Thus $S^{(1)}[z_1, z_2]$ is infinite.

If condition (1.35) holds, then, by the lemma proved above, either $S^{(1)}[z] \neq \varnothing$ for $z \in A[Z]$, or $S^{(1)}[q] \neq \varnothing$ for $q \in P[Z]$. Obviously, $S^{(1)}[q] = S^{(1)}[z_1, z_2]$ and $z_1, z_2 \in A[Z]$, if $z_1, z_2 \in A$, $\operatorname{acl}(q) = \operatorname{acl}(z_1, z_2)$. □

From this point on, let A be a locally finite infinite homogeneous semiprojective geometry; $[a_1, \dots, a_n]$ will denote $A[a_1, \dots, a_n]$, Π will be an arbitrary plane in A, i.e., $\Pi = [a_1, a_2, a_3]$ for some independent $a_1, a_2, a_3 \in A$. We shall introduce certain parameters of Π which, thanks to homogeneity, are invariants of the geometry A. We denote:

k—the number of points on an arbitrary line $[a_1, a_2]$;

s—the number of lines on Π which pass through some point $a \in \Pi$ and do not intersect a given line $[a_1, a_2] \subseteq \Pi$ such that $a \notin [a_1, a_2]$;

$k + s$—the number of lines on Π which pass through some point of Π;

p—the number of points on Π;

q—the number of lines on Π.

It is easily verified that

(1) $p = (k + s)(k - 1) + 1$,

(2) $q = \frac{p\cdot(k+s)}{k} = \frac{(k+s)^2(k-1)+k+s}{k}$.

Define $A' = \widehat{A}_{\{a\}}$ for some point $a \in A$. Denote the parameters of A' by k', s', p', q', respectively. From the definition of localization, it follows at once that

(3) $k' = k + s$.

We claim that also

(4) $s' \leq \frac{(s+k)(k-1)}{k} \cdot s$.

PROOF OF (4). Given an arbitrary quadruple of independent points a_1, a_2, a_3, b in A, consider the subspace $\Sigma = [a_1, a_2, a_3, b]$, $\Pi \subseteq \Sigma$. Let x be an arbitrary point of Σ other than b. By semiprojectivity, $x \in [b, y_1, y_2]$ for some points $y_1, y_2 \in \Pi$, or, in other words,

(5) the line $[b, x]$ is coplanar with a certain line $[y_1, y_2]$ of Π.

To find a bound for s', we look at the subspace $\widehat{\Sigma}_{\{a_1\}}$ of A'. It is obvious that $[a_1, a_2, a_3]$ is a line in the plane $\widehat{\Sigma}_{\{a_1\}}$ and does not contain the point $[a_1, b]$, while s' is the number of planes in Σ of the form $[a_1, b, x]$ such

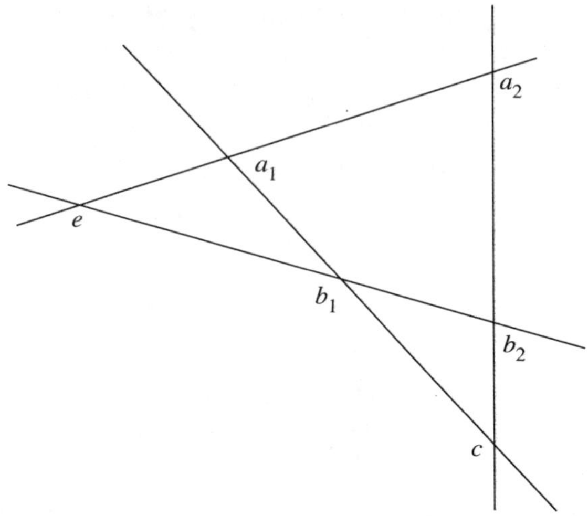

FIGURE 3.1. THE CONFIGURATION $\Phi(a_1, a_2, b_1, b_2, c, e)$.

that

$$[a_1, b, x] \cap \Pi = \{a_1\}.$$

On every plane in Σ there are exactly $s + k - 1$ lines of the form $[b, x]$ where $[b, x] \neq [b, a_1]$, and each such line determines a plane $[a_1, b, x]$; hence the number of lines $[b, x]$ such that

(6) $[b, x] \neq [b, a_1]$ and $[a_1, b, x] \cap \Pi = \{a_1\}$,

is equal to $s' \cdot (s + k - 1)$.

On the other hand, in view of (5) every line $[b, x]$ lies on a plane $[b, y_1, y_2]$, where $[y_1, y_2] \subseteq \Pi$. Obviously, if $[b, x]$ satisfies (6), then $a_1 \notin [y_1, y_2]$. Therefore, the number of planes of the form $[b, y_1, y_2]$ on which there is a line $[b, x]$ satisfying (6) is at most $q - k - s$. It also follows from (6) that $[b, x] \cap [y_1, y_2] \neq \varnothing$ if $[y_1, y_2] \subseteq \Pi$, $y_1 \neq y_2$. Therefore, on every plane $[b, y_1, y_2]$ there are at most s lines $[b, x]$ satisfying (6). Hence

$$s' \cdot (s + k - 1) \leq (q - k - s) \cdot s.$$

By transforming this inequality with the help of (2), we obtain inequality (4). □

Let a_1, a_2, b_1, b_2, c, e be distinct points lying on one plane. We let $\Phi(a_1, a_2, b_1, b_2, c, e)$ denote the following condition:

a_1, b_1, e are not collinear;

c is the intersection of the lines $[a_1, b_1]$ and $[a_2, b_2]$;

e is the intersection of the lines $[a_1, a_2]$ and $[b_1, b_2]$ (see Figure 3.1).

LEMMA 2.3. *If $k \geq 3$, then either in A or in A' there are points a_1, a_2, b_1, b_2, c, e such that $\Phi(a_1, a_2, b_1, b_2, c, e)$.*

PROOF. Suppose that $\Phi(a_1, a_2, b_1, b_2, c, e)$ is not valid in A. Consider a pair of distinct lines α, β in the plane Π, with common point e. For any distinct points $a_1, a_2 \in \alpha - \{e\}$ and $b_1, b_2 \in \beta - \{e\}$, we have $[a_1, b_1] \cap [a_2, b_2] = \varnothing$; hence the number of points lying on all possible lines of the form $[a, b]$, where $a \in \alpha - \{e\}$, $b \in \beta - \{e\}$, but not on $\alpha \cup \beta$, is exactly $(k-1)^2(k-2)$. Since $\alpha \cap \beta = \{e\}$, it follows that $\|\alpha \cup \beta\| = 2k - 1$,

(7) $p \geq (k-1)^2(k-2) + 2k - 1$.

Analogous arguments, assuming that $\Phi(a_1', a_2', b_1', b_2', c', e')$ is not valid in A', show that also

(7′) $p' \geq (k'-1)^2(k'-2) + 2k' - 1$.

Using (1), one easily sees that if $k \geq 2$ (7) is equivalent to the inequality

(8) $s \geq (k-2)^2$,

and (7′), respectively, to

(8′) $s' \geq (k'-2)^2$.

Inequality (8′) together with (4) and (3) gives

$$(s+k-2)^2 \leq \frac{(s+k)(k-1)}{k} \cdot s,$$

or, equivalently,

(9) $s(s + k^2 - 3k) + k(k-2) \leq 0$.

It is easy to see that if $k \geq 3$ inequality (8) contradicts inequality (9). □

PROPOSITION 2.4. *For any degenerate infinite locally finite homogeneous pregeometry A, there exists a finite set $X \subseteq A$ such that the geometry $\widehat{A}_X$ contains points a_1, a_2, b_1, b_2, c, e for which $\Phi(a_1, a_2, b_1, b_2, c, e)$.*

PROOF. Since A is a nondegenerate pregeometry, it follows that for some finite $X \subseteq A$ the lines of $\widehat{A}_X$ contain at least three points, and $\widehat{A}_X$ is semiprojective. Now, applying Lemma 2.3, we obtain the required conclusion. □

§3. Translation group

Throughout this section A will be a fixed $\varnothing$-definable strongly minimal set in a totally categorical structure $\mathfrak{M}$.

It should be noted that many propositions and proofs in this section are valid under assumptions weaker than total categoricity. In particular, Lemmas 3.2–3.4 require only ω-stability of $\mathfrak{M}$. Facts of this kind will be utilized in the next chapter.

An X-definable partial map $g: A \to A$ will be called a *quasitranslation* on A if $\operatorname{Dom} g \not\subseteq A[X]$, $\operatorname{Range} g \not\subseteq A[X]$.

LEMMA 3.1. *Any X_0-definable quasitranslation g on A is a bijective map of $A - A[X_0]$ onto itself.*

PROOF. Since A is strongly minimal,

$$A - A[X_0] \subseteq \operatorname{Dom} g, \qquad g(A - A[X_0]) = A - A[X_0].$$

We may assume without loss of generality that X_0 is finite. Let $X \supseteq X_0$ be finite, $x \in A - A[X_0]$. If $g(x) \in A[X] - A[X - 0]$ then, since $g(x) \in A[X_0 \cup \{x\}] - A[X_0]$, we have $x \in A[X_0 \cup \{g(x)\}] \subseteq A[X]$. Therefore, g maps $A[X] - A[X_0]$ onto itself, and moreover bijectively, as this set is finite. Hence g is bijective on $A - A[X_0]$. □

Obviously, the set of all quasitranslations on A is a semigroup relative to the operation of superposition; we denote this semigroup by QT. Define a congruence relation E on QT as follows: for $g_1, g_2 \in QT$:

$$g_1 E g_2 \quad \text{means that} \quad R(\{x: g_1(x) = g_2(x)\}) > 0.$$

It follows from Lemma 3.1 that the quotient QT/E is a group. We denote this group by T and call it the *translation group.*

LEMMA 3.2. *Let g_1, g_2 be quasitranslations which are definable relative to X_1 and X_2, respectively, and assume that g_1 equals g_2 modulo E. Then for any $a \in A - (A[X_1] \cup A[X_2])$ we have $g_1(a) = g_2(a)$.*

PROOF. Let $h(\overline{x}, v)$, with $\overline{x}$ fixed, be a quasitranslation (denoted by $h_{\overline{x}}$) equal to g_1 (or g_2) modulo E, $h(\overline{w}, v)$ a $\varnothing$-definable map. Consider the set

$$P = \{\overline{y}: h_{\overline{y}} E g_1\} = \{\overline{y}: h_{\overline{y}} E g_2\}.$$

P is X_1-definable and X_2-definable. Therefore, letting p be the complete type over $X_1 \cup X_2 \cup \{a\}$ containing P and such that $R(p) = R(P)$, and letting $\overline{x}$ be an element realizing p, we have

$$h_{\overline{x}} E g_1, \qquad h_{\overline{x}} E g_2,$$
$$R(\overline{x}, X_1 \cup X_2 \cup \{a\}) = R(\overline{x}, X_1 \cup \{a\}) = R(\overline{x}, X_1),$$
$$R(\overline{x}, X_1 \cup X_2 \cup \{a\}) = R(\overline{x}, X_2 \cup \{a\}) = R(\overline{x}, X_2).$$

Using the Reciprocity Principle, we obtain

$$a \notin A[X_1 \cup \{\overline{x}\}], \qquad a \notin A[X_2 \cup \{\overline{x}\}],$$

and consequently,

$$g_1(a) = h_{\overline{x}}(a), \qquad g_2(a) = h_{\overline{x}}(a),$$

whence follows the required equality. □

Given a finite set X, we shall say that a Y-definable quasitranslation g on A is *X-nontrivial* if, for some $a \in A - A[X \cup Y]$, the elements a, $g(a)$ are independent over X.

A quasitranslation will be called *nontrivial* if it is $\varnothing$-nontrivial.

LEMMA 3.3. *Let A be a strongly minimal set on which there exists an X-nontrivial quasitranslation. Then for any $X' \supseteq X$ there exists an X'-nontrivial quasitranslation on A.*

PROOF. Let $f(\overline{y}, v)$ be a nontrivial quasitranslation on A, where $f(\overline{w}, v)$ is an X-definable map. Choose $\overline{y}'$ so that $\operatorname{tp}(\overline{y}', X) = \operatorname{tp}(\overline{y}, X)$, $R(\overline{y}', X') =$

$R(\overline{y}, X)$. It is clear that $f(\overline{y}', v)$ defines a quasitranslation on A. We claim that this quasitranslation is X'-nontrivial. Indeed, since $f(\overline{y}, v)$ is X-nontrivial, it follows that for some $a \in A - A[X \cup |\overline{y}|]$ we have

$$R(f(\overline{y}, a), X \cup \{a\}) = 1.$$

Since $R(f(\overline{y}, a), X \cup \{a\} \cup |\overline{y}|) = 0$, it follows by the Reciprocity Principle that

$$R(\overline{y}, X \cup \{a, f(\overline{y}, a)\}) < R(\overline{y}, X \cup \{a\}) = R(\overline{y}, X).$$

Since $a \notin \mathrm{acl}(X \cup |\overline{y}|)$, there exists $a' \in A - A[X' \cup |\overline{y}'|]$ such that $\mathrm{tp}(\overline{y}'{}^\frown a', X) = \mathrm{tp}(\overline{y}{}^\frown a, X)$. Then

$$\begin{aligned} R(\overline{y}', X' \cup \{a', f(\overline{y}', a')\}) &\le R(\overline{y}', X \cup \{a', f(\overline{y}', a')\}) \\ &< R(\overline{y}', X) = R(\overline{y}', X'). \end{aligned}$$

Therefore $f(\overline{y}', a') \notin \mathrm{acl}(X' \cup \{a'\})$, for otherwise $R(\overline{y}', X' \cup \{a'\}) < R(\overline{y}', X')$, and by the Reciprocity Principle $R(a', X' \cup |\overline{y}'|) < R(a', X') = 1$, contrary to the fact that a' was chosen outside $\mathrm{acl}(X' \cup |\overline{y}'|)$. □

Lemma 3.4. *If a quasitranslation g on A is X-definable and g' is X-nontrivial, then the set $\{a \in A: g(a) = g'(a)\}$ is finite.*

Proof. For all $a \in A - A[X]$, the elements a, $g(a)$ are dependent over X and therefore, if $g'(a) = g(a)$, then $a \in [X \cup Y]$, where Y is a finite set such that g' is Y-definable. □

Lemma 3.5. *The translation group on A is infinite if and only if there exists a nontrivial quasitranslation on A.*

Proof. Suppose there exists a nontrivial quasitranslation on A. Let $g_1, \dots, g_k$ be arbitrary quasitranslations on A. They are all X-definable for some finite X. By Lemma 3.3, there exists an X-nontrivial quasitranslation g on A. By Lemma 3.2, g is distinct from all g_1 modulo E. Thus T is an infinite group.

Conversely, suppose that all quasitranslations on A are trivial. Then the graph of each of them is almost all contained in $(A \times A)^{(1)}[\varnothing]$. If there existed infinitely many distinct quasitranslations modulo E, it would follow that the rank 1 set $(A \times A)^{(1)}[\varnothing]$ contained infinitely many strongly minimal subsets having finite intersections with one another—but that is impossible. □

Lemma 3.6. *There exist a natural number* 1 *and a $\varnothing$-definable map $f: A^{l+1} \to A$ such that any map $f_{\overline{x}}: v \to f(\overline{x}, v)$, $\overline{x} \in A^l$, is a quasitranslation on A and any quasitranslation on A is equal modulo E to some $f_{\overline{x}}$.*

Proof. Consider an arbitrary X-definable quasitranslation g on A, $X \subseteq A$. By Theorem 2.1, for any $a, b \in A$, if $b \in A[X \cup \{a\}]$, there exist $x_1, x_2 \in A[X]$ such that $b \in A[a, x_1, x_2]$. Then for any $a \in A - A[X]$ there exist $x_1, x_2 \in A[X]$ such that $g(a) \in A[a, x_1, x_2]$, since $g(a) \in A[X \cup \{a\}]$.

Thus, for infinitely many elements a of A there exist $x_1, x_2 \in A[X]$ such that $g(a) \in A[a, x_1, x_2]$. Since A is strongly minimal, there exist $x_1, x_2 \in A[X]$ such that $g(a) \in A[a, x_1, x_2]$ for almost all $a \in A$, and consequently $g \subseteq (A \times A)^{(1)}[x_1, x_2]$. Since the rank of the latter set is 1, there exist only finitely many quasitranslations which are different modulo E and intersect with $(A \times A)^{(1)}[x_1, x_2]$ in an infinite set. Let these quasitranslations be $f_1(\overline{x}, v), \ldots, f_n(\overline{x}, v)$ for a suitable $\overline{x} \in A^n$ and $\varnothing$-definable maps $f_i(\overline{w}, v)$, $i = 1, \ldots, k$.

Since the number of types of pairs $\langle x_1, x_2\rangle \in A^2$ is finite, we may assume that for any quasitranslation g on A there exist $\overline{x} \in A^n$ and $i \in \{1, \ldots, k\}$ such that $f_i(\overline{x}, v)$ is a quasitranslation equal to g modulo E.

We now set $l = n + k + 1$,

$$f(\overline{x}, u_0, \ldots, u_k, v) = \begin{cases} f_i(\overline{x}, v), & \text{if } f_i(\overline{x}, v) \text{ is a quasitranslation} \\ & \text{on } A,\ u_0 = u_i,\ u_0 \neq u_j \text{ for } i \neq j; \\ v & \text{otherwise}. \end{cases}$$

LEMMA 3.7. *The translation group T is $\varnothing$-definable in $\mathfrak{M}$. Moreover, T is 2-coordinatizable over A.*

PROOF. It is easy to see that the map $\overline{x} \to f_{\overline{x}}E$ is the desired interpretation of T in $\mathfrak{M}$.

Consider the $\varnothing$-definable correspondence between T and A^2 defined by the rule: $\langle x_1, x_2\rangle$ corresponds to an element t of T if there is a quasitranslation g on A such that $t = gE$, $g(v) \in A[v, x_1, x_2]$ for almost all $v \in A$. As shown in the proof of Lemma 3.6, each $t \in T$ there corresponds a certain pair $\langle x_1, x_2\rangle$, each pair $\langle x_1, x_2\rangle$ being associated with only finitely many $t \in T$. Therefore, if $\langle x_1, x_2\rangle$ corresponds to t, then $t \in \operatorname{acl}(x_1, x_2)$ and so T is 2-coordinatizable. □

It follows from Lemma 3.2 that the action of QT on A induces a partial function $\langle t, a\rangle \to ta$ from $T \times A$ to A, defined for pairs $\langle t, a\rangle$ such that $a \notin \operatorname{acl}(t)$ and given by the rule $ta = f_{\overline{x}}(a)$, where $\overline{x} \in A^l$, $f_{\overline{x}}E = t$, $a \notin \operatorname{acl}(\overline{x})$.

LEMMA 3.8. (a) *The map $\langle t, a\rangle \to ta$ is $\varnothing$-definable in $\mathfrak{M}$.*

(b) *$t_1 = t_2$ if and only if $t_1 a = t_2 a$ for almost all $a \in A$.*

(c) *The map $a \to ta$ maps $A - A[t]$ bijectively onto itself.*

(d) *If $a \in A - A[t_1, t_2]$, then the values of $t_1(t_2 a)$ and $(t_1 \cdot t_2)a$ are defined and coincide.*

PROOF. (a) Obvious.

(b) If $t_1 a = t_2 a$ for almost all $a \in A$, then also for $a \in A - A[t_1, t_2]$. Suppose that for $i = 1, 2$ $t_i = f_{\overline{x}_i}E$, $\overline{x}_i, a$ are independent over $\{t\}$. Then $f_{\overline{x}_1}(a) = t_1 a = t_2 a = f_{\overline{x}_2}(a)$. Since $a \notin \operatorname{acl}(|\overline{x}_1| \cup |\overline{x}_2|)$, it follows that $f_{\overline{x}_1}E = f_{\overline{x}_2}E$, and so $t_1 = t_2$.

(c) Let $a_1, a_2 \in A - A[t]$, $a_1 \neq a_2$. Consider $\overline{x}$ such that $f_{\overline{x}}E = t$ and $R(\overline{x}, t) = R(\overline{x}, \{t, a_1, a_2\})$. Then $ta_1 = f_{\overline{x}}(a_1) \neq f_{\overline{x}}(a_2) = ta_2$. Thus the map $a \to ta$ is injective on $A - A[t]$. Since the image of $A - A[t]$ is infinite, the strong minimality implies that $A - A[t]$ coincides with $t(A - A[t])$.

(d) Let $a \in A - A[t_1, t_2]$. Clearly, then $a \notin \operatorname{acl}(t_1 \cdot t_2)$ and $(t_1 \cdot t_2)a$ is defined. By part (c), $t_2 a \in A - A[t_1, t_2]$; therefore $t_1(t_2 a)$ is defined. Now let $f_{\overline{x}}$, $f_{\overline{x}_1}$, $f_{\overline{x}_2}$ be representatives of $t_1 \cdot t_2, t_1, t_2$, respectively, such that $a \notin \operatorname{acl}(|\overline{x}| \cup |\overline{x}_1| \cup |\overline{x}_2|)$, $(t_1 \cdot t_2)a = f_{\overline{x}}(a)$, $t_2 a = f_{\overline{x}_2}(a)$, $t_1(t_2 a) = f_{\overline{x}_1}(t_2 a)$. Clearly, $f_{\overline{x}}E = f_{\overline{x}_1}E \cdot f_{\overline{x}_2}E$, and therefore $f_{\overline{x}}(a) = f_{\overline{x}_1}(f_{\overline{x}_2}(a))$, since $a \notin \operatorname{acl}(|\overline{x}| \cup |\overline{x}_1| \cup |\overline{x}_2|)$. But this means that $t_1(t_2 a) = (t_1 \cdot t_2)a$. □

We have thus proved

PROPOSITION 3.9. *The translation group T is $\varnothing$-definable in $\mathfrak{M}$ together with a partial action $\langle t, a\rangle \to ta$ of T on the set A. Moreover, the map $a \to ta$ is a bijection of $A - A[t]$ onto itself for any $t \in T$. If $t = gE$, then $ta = g(a)$ for almost all $a \in A$. The group T is infinite if and only if there exists a nontrivial quasitranslation of A, and in that event $R(T) \leq 2$.*

Let us assume now that for some group G

(1) G is $\varnothing$-definable in $\mathfrak{M}$, $RD(G) = \langle 2, 1\rangle$, Q is a strongly minimal $\varnothing$-definable subset of G which is an atom and satisfies the following condition:

(2) for any strongly minimal subgroup H of G and any $g \in G$, $g \cdot H \cap Q$ is finite.

It follows from condition (2) that the subgroup

$$H_0 = \{h \in G\colon R(h \cdot Q \cap Q) = 1\}$$

is finite. Indeed, if H_0 were infinite then, as follows from Fact 1.4.6, H_0 would be strongly minimal and $H_0 \cdot g \cap Q$ infinite. But then $g \cdot H \cap Q$, $H = g^{-1} \cdot H_0 \cdot g$, would be infinite, contrary to (2).

Obviously, H_0 is a $\varnothing$-definable subgroup, and therefore the set

$$Q_0 = \bigcap \{h \cdot Q\colon h \in H_0\}$$

is also $\varnothing$-definable. Since Q_0 is not empty, $Q_0 \subseteq Q$ and Q is an atom, we have $Q_0 = Q$. We have thus shown that $g \cdot Q \cap g' \cdot Q$ is infinite if and only if $g \cdot Q = g' \cdot Q$ and $g' \in g \cdot H_0$. Now consider the incidence system $\mathscr{P}(G, Q) = (P, L, I)$, where $P = G - G^{(1)}[\varnothing]$, $L = \{g \cdot H_0\colon g \in G - G^{(1)}[\varnothing]\}$ and for $p \in P$, $l = g \cdot H_0 \in L$, the condition pIl stands for $p \in g \cdot Q$.

Since H_0 is a finite subgroup, it follows that $Q \cdot Q$ is a set of rank 2 in G, i.e., $RD(G) = \langle 2, 1\rangle$. Since $G^{(1)}[\varnothing]$ is a maximal $\varnothing$-definable subset of G and $D(G) = 1$, it follows that P and L are atoms. In addition, $G - G^{(1)}[\varnothing] \subseteq Q \cdot Q$, whence it follows that P and L are 2-coordinatizable over Q, and by considerations of rank this coordinatization is strict. We have thus proved

LEMMA 3.10. *If conditions* (1) *and* (2) *are fulfilled,* $\mathscr{P}(G, Q)$ *is a quasidesign over* Q.

LEMMA 3.11. *Let conditions* (1) *and* (2) *be fulfilled, and let* B *be a definable strongly minimal subset of* G *such that* $B \cap g \cdot Q$ *is finite for all* $g \in G$. *Then there exists a strongly minimal subgroup* H *of* G *such that* $H \subseteq G^{(1)}[\varnothing]$ *and* $B \cap g \cdot H$ *is infinite for some* $g \in G$.

PROOF. Let B be definable over X. Since $R(G^{(1)}[\varnothing]) = 1$, the set

$$\{g \in G \colon R(g \cdot B \cap G^{(1)}[\varnothing]) = 1$$

is partitioned by the equivalence relation " $g_1 B \cap g_2 B$ is infinite" into finitely many classes. By Fact 1.4.6, the rank of each such class is ≤ 1. Therefore, if $g \in G$, $R(g, X) = 2$, then all but a finite number of the elements of $g \cdot B$ lie in the set $P = G - G^{(1)}[\varnothing]$. By Proposition 1.4, which is applicable to the quasidesign $\mathscr{P}(G, Q)$, in view of the fact that by assumption $g \cdot B \cap h \cdot Q$ is finite for any $h \in G$, we see that $g \cdot B$ satisfies condition (1.35). Let $g_0, g_1 \in B$ be independent over $X \cup \{g\}$. Then $g \cdot g_0$, $g \cdot g_1$ are also independent over $X \cup \{g\}$ and therefore, by (1.35),

$$R(g \cdot g_1, \{g \cdot g_0\}) = 1.$$

This is equivalent to the condition

$$R(g_0^{-1} \cdot g_1, \{g \cdot g_0\}) = 1,$$

since $g_0^{-1} \cdot g_1 = (g \cdot g_0)^{-1} \cdot g \cdot g_1$, $\operatorname{acl}(g \cdot g_1, g \cdot g_0) = \operatorname{acl}(g_0^{-1} \cdot g_1, g \cdot g_0)$.

If we now assume that $R(g_0^{-1} \cdot g_1, \varnothing) = 2$, then the elements $g_0^{-1} \cdot g_1$ and $g \cdot g_0$ are dependent, i.e.,

$$R(g \cdot g_0, \{g_0^{-1} \cdot g_1\}) \leq 1,$$

and hence

$$R(g \cdot g_0, \{g_0, g_1\}) \leq 1 \quad \text{and} \quad R(g, \{g_0, g_1\}) \leq 1.$$

This contradicts the choice of the elements g, g_0, g_1. Consequently, $g_0^{-1} \cdot g_1 \in G^{(1)}[\varnothing]$. Thus, the map

$$\pi \colon B \times B \to G^{(1)}[\varnothing], \quad \pi(g_0, g_1) = g_0^{-1} \cdot g_1,$$

is defined on all pairs g_0, g_1 that are independent over X. Since $R(\operatorname{Dom} \pi) = 2$ but $R(\operatorname{Range} \pi) \leq 1$, it follows that for any pair g_0, g_1 of independent elements of B there exists an infinite set of pairs $\langle h_0, h_1 \rangle \in \operatorname{Dom} \pi$ such that $g_0^{-1} \cdot g_1 = h_0^{-1} \cdot h_1$. Hence it follows that if $h_0 \in B - B[\{g_0, g_1\} \cup X]$, there exists h_1 such that $h_0^{-1} \cdot h_1 = g_0^{-1} \cdot g_1$. Since $B \cdot h$ contains h_1 if $h = h_0^{-1} \cdot h_1$ and $h_0, h_1 \in B$, it follows from the previous arguments that

$B \cdot h \cap B$ is infinite for $h = g_0^{-1} \cdot g_1$, where g_0, g_1 are elements of B independent over X. Thus the group

$$H = \{h \in G\colon R(B \cdot h \cap B) = 1\}$$

is infinite, and in fact $H \cap G^{(1)}[\varnothing]$ is infinite.

By 1.4.6, H is strongly minimal and for some $g \in B$ $gH \cap B$ is infinite. Since $H - G^{(1)}[\varnothing]$ is finite and there are only finitely many such subgroups, it follows that $H - G^{(1)}[\varnothing] \subseteq \mathrm{acl}(\varnothing)$, and therefore $H \subseteq G^{(1)}[\varnothing]$. □

PROPOSITION 3.12. *If the group T is infinite, i.e., there exists a nontrivial quasitranslation, then T contains a strongly minimal subgroup.*

PROOF. If $R(T) = 1$, the connected component of T is the required strongly minimal subgroup.

Let $R(T) = 2$ and let G be the connected component. Enlarging the signature if necessary by the addition of constants, we may assume that G contains a strongly minimal $\varnothing$-definable subset which is an atom.

Suppose now that G contains no strongly minimal subgroup. Then conditions (1) and (2) are fulfilled.

Let $a \in A - A[\varnothing]$ and $G_a = \{t \in G\colon R(t, \{a\}) = 2\}$. Obviously, $G_a \subseteq G - G^{(1)}[\varnothing]$ and $a \notin \mathrm{acl}(t)$ for all $t \in G_a$. For $b \in A$, we denote

$$G_a^b = \{t \in G_a\colon ta = b\}.$$

Since

$$R(G_a) = 2, \qquad G_a = \bigcup\{G_a^b\ b \in A - A[\varnothing]\},$$

there exists $b \in A - A[\varnothing]$ such that

$$R(G_a^b) \geq 1.$$

Let B be a strongly minimal subset of G_a^b. Under our assumption, it follows from Lemma 3.11 that $B \cap t \cdot Q$ is infinite for some $t \in G$. Hence it follows that $(t \cdot q)a = b$ for all but finitely many elements q of Q. Let $s \in G$, $R(s, \{t, a, b\}) = 2$. Then $R(s \cdot t, \{t, a, b\}) = 2$. If $q \in Q - Q[t, a]$ and s, q are independent over $\{t, a\}$, then $a \notin \mathrm{acl}(s, t\cdot q)$ and $s((t\cdot q)a) = (s \cdot t \cdot q)a$, by 3.8(d). On the other hand, by assumption $a \notin \mathrm{acl}(s \cdot t, q)$ and therefore $(s \cdot t \cdot q)a = (s \cdot t)(qa) = sb$.

This implies that $qa = c$ for some fixed $c \in A$ and for all $q \in Q - Q[a, c]$. Hence $qa = q'a$ for all $q, q' \in Q$ such that $a \notin \mathrm{acl}(q, q')$. But then $q = q'$ by 3.8(b), contradicting the assumption that Q is infinite. Consequently, G contains a strongly minimal subgroup. □

§4. Construction of a nontrivial quasidesign

We shall say that a nontrivial quasidesign is definable in a structure $\mathfrak{M}$ if, in some enrichment $\mathfrak{M}'$ of $\mathfrak{M}$ by constant symbols, there is a $\varnothing$-definable strongly minimal set A on which a nontrivial quasidesign is definable. In this section $\mathfrak{M}$ will again be a totally categorical structure.

PROPOSITION 4.1. *If a strongly minimal set whose geometry is not degenerate is definable in* $\mathfrak{M}$, *then a nontrivial quasidesign is definable in* $\mathfrak{M}$.

The proof of this proposition will take up this entire section.

Proceeding to the proof of the proposition, we observe that, by Proposition 2.4, we may consider a geometry A, definable in $\mathfrak{M}$, in which the condition $\Phi(a_1, a_2, b_1, b_2, c, e)$ is valid.

Fix $e \in A$ and denote

$$U^e = \{\langle a, b\rangle \in A^2 : a \in [b, e],\ a \neq b\}.$$

Clearly, U^e is definable relative to $\{e\}$ and $R(U^e) = 1$. By Shelah's Finite Equivalence Relation Theorem (Fact 1.3.6), there exists an e-definable equivalence relation E on U^e such that every E-class is strongly minimal or finite. Since the union of all the finite nonempty classes is e-definable and finite, it lies in $U^e[e]$. Since the latter, as is easily seen, is empty, it follows that all the E-classes are strongly minimal.

Let

(1) $U^e = \bigcup\{U_i^e : 0 \leq i \leq m\}$ be the partition induced by this equivalence relation.

Given $d \in A - \{e\}$, $0 \leq i \leq m$, let us denote

$$u_i = \|\{b \in A : \langle d, b\rangle \in U_i^e\}\|.$$

It follows from homogeneity that the u_i's are independent of the choice of d, e. Moreover, we have the following

LEMMA 4.2. *Let* $\langle a_1, a_2\rangle \in U_i^e$ *and* $\langle a_1', a_2'\rangle \in U_i^e$, $b \in A - ([a_1, e] \cup [a_1', e])$. *Then* $\mathrm{tp}(\langle a_1, a_2\rangle, [b, e]) = \mathrm{tp}(\langle a_1', a_2'\rangle, [b, e])$.

PROOF. Since E is e-definable, the equivalence class U_i^e is definable by any of its representatives together with e. Since $a_1', b \in A - [a_1, a_2, e]$ by homogeneity, there exists $b' \in [b, e]$ such that $\langle b, b'\rangle \in U_i^e$.

The equivalence class U_i^e is definable by its representative $\langle b, b'\rangle$, that is, U_i^e is $[b, e]$-definable. The required equality is now derived from the strong minimality of U_i^e, bearing in mind that $\langle a_1, a_2\rangle$, $\langle a_1', a_2'\rangle \in U_i^e - \mathrm{acl}(b, e)$. □

Denote

$$V^e(b_1, b_2, c) = \{\langle a_1, a_2\rangle \in U^e : \Phi(a_1, a_2, b_1, b_2, c, e)\},$$
$$V_i^e(b_1, b_2, c) = V^e(b_1, b_2, c) \cap U_i^e, \qquad v_{ij} = \|V_i^e(b_1, b_2, c)\|$$

for some $b_1, b_2, c \in A$ such that $\langle b_1, b_2\rangle \in U_j^e$ and $c \notin [b_1, b_2]$. By homogeneity, v_{ij} is independent of the choice of b_1, b_2, c.

LEMMA 4.3. *If* $v_{ij} \neq 0$, *then* $v_{ij} = u_i$.

PROOF. Let Π be a fixed plane in A, $e \in \Pi$, $\langle b_1, b_2\rangle \in U_j^e$, $[b_1, b_2] \subseteq \Pi$, $v_{ij} \neq 0$. Let us calculate the number of elements in the set

$$S_i^e(b_1, b_2) = \{\langle a_1, a_2\rangle \in U_i^e : a_1, a_2 \in \Pi - [b_1, b_2]\};$$
$$\|S_i^e(b_1, b_2)\| = (p-k)\cdot u_i.$$

On the other hand, it follows from Lemma 4.2 and the fact that $v_{ij} \neq 0$ that for any $\langle a_1, a_2\rangle \in U_i^e$ such that $b_1 \notin [a_1, e]$ there exists $c \in \Pi - [b_1, b_2]$ such that $\Phi(a_1, a_2, b_1, b_2, c, e)$.

Hence
$$S_i^e(b_1, b_2) = \bigcup\{V_i^e(b_1, b_2, c) : c \in \Pi - [b_1, b_2]\},$$
that is,
$$\|S_i^e(b_1, b_2)\| = v_{ij}\cdot(p-k),$$
which proves the lemma. □

LEMMA 4.4. *Let* $\langle a_1, a_2\rangle \in U_i^e$, $u_i = 1$ *and* $\Phi(a_1, a_2, b_1, b_2, c, e)$ *for some* b_1, b_2, c. *Then a nontrivial quasidesign is definable in* $\mathfrak{M}$.

PROOF. Let $\langle b_1, b_2\rangle \in U_j^e$. We shall show that a nontrivial quasidesign g is definable on U_i^e.

Define $g(\langle a_1, a_2\rangle)$ as the pair $\langle c, c'\rangle \in U_i^e$, where $\Phi(a_1, a_2, b_1, b_2, c, e)$, c' is the unique element (since $u_i = 1$) such that $\langle c, c'\rangle \in U_i^e$.

We now choose $d \in A - [a_1, b_1, e]$ (note that $[a_1, b_1, e] = [a_1, a_2, b_1, b_2, c, c', e]$). Since the equivalence classes U_i^e are definable by any of their representatives, it follows that U_i^e is $[d, e]$-definable. It easily follows from the definition that

$$\langle a_1, a_2\rangle \notin \operatorname{acl}(b_1, b_2, d, e), \qquad \langle c, c'\rangle \notin \operatorname{acl}(a_1, a_2, d, e).$$

We can obviously find elements in U_i^e of the form $\langle b_1, b_1'\rangle$, $\langle b_2, b_2'\rangle$, where b_1', b_2' are certain elements of A, $b_1', \in [b_1, e]$, $b_2' \in [b_2, e]$. Add $[d, e]$ to the signature and define $X = \{\langle b_1, b_1'\rangle, \langle b_2, b_2'\rangle\}$. Obviously, g is X-definable and $X \subseteq \operatorname{acl}(b_1, b_2)$. It follows from the previous observation that

$$\langle a_1, a_2\rangle \notin \operatorname{acl}(X), \qquad \langle c, c'\rangle \notin \operatorname{acl}(\langle a_1, a_2\rangle).$$

We have thus verified the conditions defining a nontrivial quasidesign. □

In view of this lemma, we may assume henceforth, throughout this section, that

(2) if $\Phi(a_1, a_2, b_1, b_2, c, e)$ for some $a_1, a_2, b_1, b_2, c \in A$ and $\langle a_1, a_2\rangle \in U_i^e$, then $u_i > 1$.

LEMMA 4.5. *Let* $\langle a_1, a_2\rangle \in U_i^e$ *and suppose that for some* b_1, b_2, c *we have* $\Phi(a_1, a_2, b_1, b_2, c, e)$. *Then* $\langle b_1, b_2\rangle \in U_i^e$ *and*

$$V_i^e(a_1, a_2, c) = V^e(a_1, a_2, c), \qquad v_{ii} = u_i.$$

Proof. Let $\langle b_1, b_2\rangle \in U_j^e$. Then $v_{ij} \neq 0$ and therefore, by Lemma 4.3 and condition (2), $v_{ij} = u_i > 1$. This means that for some $\langle a_i', a_2'\rangle \in U_i^e - \{\langle a_1, a_2\rangle\}$ it is also true that $\Phi(a_1', a_2', b_1, b_2, c, e)$. It is easy to see that then $\Phi(a_1', a_2', a_1, a_2, c, e)$. Hence it follows that $V_i^e(a_1, a_2, c) \neq \varnothing$, i.e., $v_{ii} \neq 0$, and therefore, by Lemma 4.3, $v_{ii} = u_i$.

On the other hand, if $i \neq j$, then

$$V_i^e(a_1, a_2, c) = V_i^e(b_1, b_2, c) - \{\langle a_1, a_2\rangle\},$$

that is, $v_{ii} = v_{ij} - i = u_i - 1$. Contradiction. Consequently, $i = j$. □

Let I^e be the set of all indexes i such that there exist $\langle a_1, a_2\rangle, \langle b_1, b_2\rangle \in U_i^e$ and $c \in A$ for which $\Phi(a_1, a_2, b_1, b_2, c, e)$. It follows from the lemma proved above that $I^e \neq \varnothing$. We shall assume that

(3) $0 \in I^e$ and $u_0 = \max\{u_i : i \in I^e\}$.

Let α, β be two different lines in A and e their common point, $c \in A - (\alpha \cup \beta)$. Denote

$$N(\alpha, \beta, c) = \{a \in \alpha : (\exists b \in \beta)([a, b] = [a, c] \,\&\, a \neq e)\},$$
$$U_0^e(\alpha, \beta, c) = \{\langle a_1, a_2\rangle \in U_0^e : [a_1, a_2] = \alpha \,\&$$
$$(\exists b_1, b_2 \in \beta)\Phi(a_1, a_2, b_1, b_2, c, e)\}.$$

Lemma 4.6. *If* $\|U_0^e(\alpha, \beta, c)\| \geq u_0^2$, *then* $\|N(\alpha, \beta, c)\| = u_0 + 1$ *and*

(4) $U_0^e(\alpha, \beta, c) = \{\langle a_1, a_2\rangle \in N(\alpha, \beta, c)^2 : a_1 \neq a_2\}$.

Proof. It is evident from the definition that the left-hand side of (4) is always a subset of the right-hand side. In addition, for any pair of points $a_1 \in N(\alpha, \beta, c)$, $b_1 \in [a_1, c] \cap \beta$,

$$N(\alpha, \beta, c) = \{a_1\} \cup \{a_2 \in A : (\exists b_2)(\langle a_2, b_2\rangle \in V^e(a_1, b_1, c))\}.$$

Hence, by (3) and Lemma 4.5, $\|N(\alpha, \beta, c)\| \leq u_0 + 1$.

It follows from our previous observations that if $\|U_0^e(\alpha, \beta, c)\| \geq u_0^2$, then for some $a_1 \in N(\alpha, \beta, c)$ and $b_1 \in [a_1, c] \cap \beta$,

$$\|\{a_2 \in N(\alpha, \beta, c) : \langle a_1, a_2\rangle \in U_0^e\}\| = u_0;$$

in particular, $\|N(\alpha, \beta.c)\| = u_0 + 1$.

Let $a_2 \in N(\alpha, \beta, c)$, $b_2 \in [a_2, c] \cap \beta$. Then $\Phi(a_1, a_2, b_1, b_2, c, e)$. By Lemmas 4.2 and 4.5,

$$\mathrm{tp}(\langle a_1, b_1\rangle, [e, c]) = \mathrm{tp}(\langle a_2, b_2\rangle, [e, c]).$$

Thus for the pair $\langle a_2, b_2\rangle$ there also exist u_0 pairs $\langle a_3, b_3\rangle$ such that $\langle a_2, a_3\rangle \in U_0^e \,\&\, \Phi(a_2, a_3, b_2, b_3, c, e)$. Since $\|N(\alpha, \beta, c)\| = u_0 + 1$, this implies (4). □

Fix e and set $U_0^e = U_0$. We shall assume throughout the sequel that if $\bar{a} \in U_0$, then $\bar{a} = \langle a_1, a_2\rangle$. If $\bar{a}, \bar{b} \in U_0$, $[a_1, a_2] \neq [b_1, b_2]$, then it follows from the condition $0 \in I^e$ and the conditions of homogeneity and strong minimality that there exists $c \in A$ for which

$$\Phi(a_1, a_2, b_1, b_2, c, e).$$

We shall write this fact as $\overline{a} * \overline{b} = c$. Clearly, c is uniquely determined by the pair $\overline{a}$, $\overline{b}$.

Lemma 4.7. *If $\overline{a}, \overline{b}, \overline{x}, \overline{x}' \in U_0$, $[a_1, a_2] \neq [b_1, b_2]$, $\overline{x} \notin \operatorname{acl}(\overline{a}, \overline{b})$, $\overline{a} * \overline{x} = \overline{a} * \overline{x}'$, $\overline{b} * \overline{x} = \overline{b} * \overline{x}'$, then $\overline{x} = \overline{x}'$.*

Let $\overline{a} * \overline{x} = c$, $\overline{b} * \overline{x} = d$. Obviously, $[x_1, x_2]$ and $[x_1', x_2']$ are subsets of the set

$$[a_1, a_2, c] \cap [b_1, b_2, d].$$

The fact that $[x_1, x_2] \not\subseteq \operatorname{acl}(\overline{a}, \overline{b})$ implies that

$$[a_1, a_2, c] \neq [b_1, b_2, d],$$

and therefore it follows from dimensionality arguments that

$$[a_1, a_2, c] \cap [b_1, b_2, d] = [x_1, x_2]$$

and

$$[a_1, a_2, c] \cap [b_1, b_2, d] = [x_1', x_2'];$$

in particular, $[x_1, x_2] = [x_1', x_2']$.

The fact that $x_1, x_1' \in [a_1, c] \cap [x_1, x_2]$ implies that $x_1 = x_1'$. Similarly, $x_2 = x_2'$, i.e., $\overline{x} = \overline{x}'$. □

Lemma 4.8. *Let $\overline{a}, \overline{b}, \overline{a}', \overline{b}', \overline{x}, \overline{y} \in U_0$, $\overline{y} \notin \operatorname{acl}(\overline{a}, \overline{b}, \overline{a}', \overline{b}')$, $[b_1, b_2] \neq [b_1', b_2']$, $[x_1, x_2] \neq [y_1, y_2]$ and*

(5) $\overline{a} * \overline{x} = \overline{b} * \overline{y}$, $\overline{a}' * \overline{x} = \overline{b}' * \overline{y}$;

then a nontrivial quasitranslation is definable on U_0.

Proof. We observe at once that, by Lemma 4.7, the element $\overline{y}$ figuring in the statement of the lemma is uniquely determined by the other parameters. Therefore, there is a partial map g, defined relative to $\{\overline{a}, \overline{b}, \overline{a}', \overline{b}'\}$, such that $g(\overline{x}) = \overline{y}$. Since $\overline{y} \notin \operatorname{acl}(\overline{a}, \overline{b}, \overline{a}', \overline{b}')$, it follows by Lemma 3.1 that we may assume g to be a quasitranslation. Since $\overline{y} \notin \operatorname{acl}(\overline{x})$, g is nontrivial. □

Define

$$P = \{\langle c_1, c_2\rangle \in A^2 : [c_1, e] \neq [c_2, e],\ c_1 \neq e,\ c_2 \neq e\},$$
$$L = \{\langle \overline{a}, \overline{b}\rangle \in U_0^2 : [a_1, a_2] \neq [b_1, b_2]\},$$
$$l(\overline{a}, \overline{b}) = \{\langle c_1, c_2\rangle \in P : (\exists \overline{x} \in U_0 - U_0[\overline{a}, \overline{b}])(c_1 = \overline{a} * \overline{x} \,\&\, c_2 = \overline{b} * \overline{x})\}.$$

Lemma 4.9. *For any $\langle \overline{a}, \overline{b}\rangle \in L$, the set $l(\overline{a}, \overline{b})$ is strongly minimal. Moreover, the map $\overline{x} \to \langle \overline{a} * \overline{x}, \overline{b} * \overline{x}\rangle$ sets up a one-to-one correspondence between $U_0 - U_0[\overline{a}, \overline{b}]$ and $l(\overline{a}, \overline{b})$.*

PROOF. That the map in question is one-to-one follows from Lemma 4.7. Since U_0 is strongly minimal, this implies that $l(\overline{a}, \overline{b})$ is strongly minimal. □

LEMMA 4.10. *Let $l(\overline{a}, \overline{b}) \cap l(\overline{a}', \overline{b}')$ be infinite for some $\langle \overline{a}, \overline{b} \rangle$, $\langle \overline{a}', \overline{b}' \rangle \in L$. Then either there is a nontrivial quasitranslation definable in $\mathfrak{M}$, or $[a_1, a_2] = [a_1', a_2']$, $[b_1, b_2] = [b_1', b_2']$ and for any $\overline{x} \in U_0 - U_0[\overline{a}, \overline{b}]$ there exists $\overline{y} \in U_0[\overline{x}]$ such that*

$$\overline{a} * \overline{x} = \overline{a}' * \overline{y}, \qquad \overline{b} * \overline{x} = \overline{b}' * \overline{y}.$$

PROOF. By assumption, there exist $\overline{x}, \overline{y} \notin U_0[\overline{a}, \overline{b}, \overline{a}', \overline{b}']$ such that $\overline{a} * \overline{x} = \overline{a}' * \overline{y}$, $\overline{b} * \overline{x} = \overline{b}' * \overline{y}$. By Lemma 4.8, we now see that either a nontrivial quasitranslation is definable, or $[x_1, x_2] = [y_1, y_2]$. In the latter case,

$$\operatorname{acl}(\overline{a}, \overline{x}) = \operatorname{acl}(\overline{a}, \overline{y}) = \operatorname{acl}(\overline{a}', \overline{x}) = \operatorname{acl}(\overline{a}', \overline{y}).$$

Hence, since $\overline{x} \notin \operatorname{acl}(\overline{a}, \overline{a}')$,

$$\operatorname{acl}(\overline{a}) = \operatorname{acl}(\overline{a}'),$$

which is equivalent to

$$[a_1, a_2] = [a_1', a_2'].$$

Similarly one shows that $[b_1, b_2] = [b_1', b_2']$. □

Define $\langle \overline{a}, \overline{b} \rangle E \langle \overline{a}', \overline{b}' \rangle$ if $l(\overline{a}, \overline{b}) \cap l(\overline{a}', \overline{b}')$ is infinite.

If no nontrivial quasitranslation is definable in $\mathfrak{M}$, it follows from Lemmas 4.9 and 4.10 that

(6) $\langle \overline{a}, \overline{b} \rangle E \langle \overline{a}', \overline{b}' \rangle \Leftrightarrow l(\overline{a}, \overline{b}) = l(\overline{a}', \overline{b}')$.

LEMMA 4.11. *Suppose that no nontrivial quasitranslation is definable in $\mathfrak{M}$. Then for $\langle \overline{a}, \overline{b} \rangle \in L$ the equivalence class $\langle \overline{a}, \overline{b} \rangle E$ includes either less than u_0^2 elements or exactly $u_0(u_0 + 1)$.*

PROOF. Fix $\langle \overline{a}, \overline{b} \rangle \in L$ and $\overline{x} \in U_0 - U_0[\overline{a}, \overline{b}]$. Let $\langle \overline{a}, \overline{b} \rangle E \langle \overline{a}', \overline{b}' \rangle$. By Lemma 4.10, $[a_1, a_2] = [a_1', a_2']$, $[b_1, b_2] = [b_1', b_2']$, and there exists $\overline{y} \in U_0[\overline{a}]$ such that

(7) $\overline{a} * \overline{x} = \overline{a}' * \overline{y} = c_1$, $\overline{b} * \overline{x} = \overline{b}' * \overline{y} = c_2$.

Note that c_1, c_2 are fixed following $\overline{a}$, $\overline{b}$, $\overline{x}$ (are independent of $\overline{a}', \overline{b}'$). Consider the lines $\alpha = [a_1, a_2]$, $\beta = [b_1, b_2]$, $\gamma = [x_1, x_2]$. It is easy to see that (7) sets up a one-to-one correspondence $\overline{y} \leftrightarrow \langle \overline{a}', \overline{b}' \rangle$ between

$$U_0^e(\gamma, \alpha, c_1) \cap U_0^e(\gamma, \beta, c_2) \quad \text{and} \quad \langle \overline{a}, \overline{b} \rangle E.$$

Therefore, if $\|\langle \overline{a}, \overline{b} \rangle E\| \geq u_0^2$, then by Lemma 4.6

$$\begin{aligned} &U_0^e(\gamma, d, c_1) \cap U_0^e(\gamma, \beta, c_2) \\ &\quad = \{\langle z_1, z_2 \rangle : z_1 \neq z_2, \ z_1, z_2 \in N(\gamma, \alpha, c_1) \cap N(\gamma, \beta, c_2)\}. \end{aligned}$$

The cardinality of the latter set may exceed $u_0^2 - 1$ only if $N(\gamma, \alpha, c_1) = N(\gamma, \beta, c_2)$, in which case it equals $u_0(u_0 + 1)$. □

LEMMA 4.12. *If no nontrivial quasitranslation is definable in $\mathfrak{M}$, then there exists a quasidesign in $\mathfrak{M}$ with parameters*

$$\pi(x) \square x^2, \quad \lambda(x) \square \frac{u_0}{u_0+1}x^2, \quad \rho(x) \square \frac{u_0^2}{u_0+1}x, \quad \kappa(x) \square u_0 x. \quad (8)$$

PROOF. Define $L^* = L/E$ and consider the structure $\mathscr{P} = (P, L, I)$, where for $p = \langle c_1, c_2 \rangle \in P$, $\langle \overline{a}, \overline{b} \rangle \in L$, $l = \langle \overline{a}, \overline{b} \rangle E \in L^*$, pIl holds if and only if $p \in l(\overline{a}, \overline{b})$. Conditions (1.1), (1.2), (1.3) are clearly satisfied for this structure.

Note that since U_0^e is X_0-definable for some X_0, $X_0 \neq \varnothing$, it follows that P and L^* need not satisfy condition (1.4) in general. We therefore consider

$$P' = \{p \in P : R(p, X_0) = 2\}, \qquad L' = \{l \in L^* : R(l, X_0) = 2\}$$

and the structure $\mathscr{P}' = (P', L', I)$. It is easy to see that $\mathscr{P}'$ inherits conditions (1.1), (1.2), (1.3) from $\mathscr{P}$. Moreover, $\pi \square \pi'$, $\lambda \square \lambda'$, $\kappa \square \kappa'$, $\rho \square \rho'$ for the parameters of these structures. But after X_0 is added to the signature it is obvious that $\mathscr{P}'$ is a quasidesign over A.

Let X be a closed subset of $\mathfrak{M}$, $X \supseteq X_0$, $x = \|A[X]\|$. Obviously,

$$\pi(x) = \|P[X]\| = (x-1)(x-k) \square x^2,$$
$$\kappa(x) = \|l(\overline{a}, \overline{b})[X]\| = \|U_0[X] - U_0[\overline{a}, \overline{b}]\| \square u_0 \cdot x \quad \text{for } \langle \overline{a}, \overline{b} \rangle \in L[X].$$

Denote $m = \|\langle \overline{a}, \overline{b} \rangle E\|$ for $\langle \overline{a}, \overline{b} \rangle \in L$. Then since $\|U_0[x]\| \square u_0 \cdot x$, it follows that

$$\lambda(x) = \|L^*[X]\| = \frac{1}{m}\|L[X]\| \square \frac{u_0^2}{m}x^2.$$

By Lemma 4.11, either $m = u_0(u_0+1)$ or $m < u_0^2$, and then $\lambda \sqsubset \pi$. The second alternative is impossible, in view of Proposition 1.3. Thus, $m = u_0(u_0+1)$, $\lambda(x) \square u_0 x^2/(u_0+1)$. Now $\rho(x) \square u_0^2 x/(u_0+1)$ by formula (1.8). □

LEMMA 4.13. *There exists no quasidesign with parameters satisfying* (8) *with* $u_0 > 1$.

PROOF. As shown in §1 (1.15),

$$\rho \cdot \kappa \square \mu_1 \cdot \sigma_1 \quad \text{and} \quad \sigma_1 \square \pi,$$

where μ_1 is an integer. In our case,

$$\rho \cdot \kappa \square \frac{u_0^3}{u_0+1}x^2, \qquad \sigma_1 \square x^2,$$

and therefore

$$\mu_1 = \frac{u_0+1}{u_0^3},$$

which is impossible for integers $u_0 > 1$. □

The last two lemmas, together with Lemma 4.4, complete the proof of Proposition 4.1.

§5. A strongly minimal group as a vector space

The key result of this section is

PROPOSITION 5.1. *If a strongly minimal set whose geometry is not degenerate is definable in $\mathfrak{M}$, then in some enrichment $\mathfrak{M}'$ of $\mathfrak{M}$ by constants there is a $\emptyset$-definable infinite vector space V over a finite field K, such that any predicate on V which is $\emptyset$-definable in $\mathfrak{M}'$ is a Boolean combination of predicates of the form*

$$k_1 v_1 + \cdots + k_n v_n + a = 0, \quad \text{where } k_1, \ldots, k_n \in K, \ a \in V[\emptyset].$$

This proposition is slightly stronger than the Classification Theorem 0.1; to be more precise, the latter follows easily from 5.1, the Trichotomy Theorem 2.3.5 and Fact 2.1.2. Indeed, putting $\mathfrak{M} = \mathfrak{A}$, we deduce from 5.1 that either the geometry $\widehat{A}$ is degenerate or there is a definable strongly minimal V such that $\widehat{V}$ is a projective geometry over K. In the latter case $\widehat{A}$ is locally projective, by 2.3.5. But local projectivity together with the fact that A is locally finite yields, by 2.1.2: $\widehat{A}$ is an affine or projective geometry over a finite field.

We proceed to the proof of Proposition 5.1.

We first observe that by 3.12 and 4.1 there is a strongly minimal group H which is definable in $\mathfrak{M}$. By 1.4.2, H is an abelian group. We may assume that H is $\emptyset$-definable.

Set $G = H \times H$ and suppose that G contains a strongly minimal subset Q satisfying condition (3.2). We may assume that Q is $\emptyset$-definable and is an atom. Then $\mathscr{P}(G, Q)$ is a quasidesign over Q, by 3.10.

LEMMA 5.2. *If conditions* (3.1) *and* (3.2) *are fulfilled, then for any finite X*

$$G^{(1)}[X] = G[X] \cdot (Q \cup U),$$

where U is the union of all strongly minimal subgroups of G contained in $G^{(1)}[\emptyset]$.

PROOF. The right-hand side is contained in the left, because

$$Q \cup U \subseteq G^{(1)}[\emptyset].$$

Let A be a strongly minimal subset of the set

$$G^{(1)}[X] - G[X] \cdot (Q \cup U).$$

If $g \cdot Q \cap A$ is infinite for some $g \in G$, then $g \cdot Q \cap G^{(1)}[X]$ is infinite. But there are only finitely many such elements g, since $R(G^{(1)}[X]) = 1$, and therefore $g \in G[X]$, contrary to the construction of A. Therefore, by Lemma 3.11, $A \cap g \cdot S$ is infinite for some strongly minimal subgroup S,

$S \subseteq U$. The same is true of $G^{(1)}[X] \cap g \cdot S$. We now show that we can choose $g \in G[X]$—this will contradict the choice of A.

By symmetry, we may assume that $S \neq \{1\} \times H = H_1$. Then, since $R(S \cdot H_1) = 2$, it follows that $S \cdot H_1 = G$ and $g \cdot S \cap H_1 \neq \varnothing$. This last intersection is finite and there exist only finitely many cosets $g' \cdot S'$, where $g' \cdot S' \cap G^{(1)}[X]$ is infinite, $S' \subseteq U$, $S' \neq H_1$, S' a strongly minimal group. The union B of all such cosets is X-definable since it is invariant under all automorphisms over X, $B \cap H_1$ is finite and $g \cdot S \cap H_1 \subseteq B \cap H_1$. Since H_1 is $\varnothing$-definable, it follows that $B \cap H_1 \subseteq \operatorname{acl}(X)$, and therefore the representative of the coset $g \cdot S$ may be chosen in $G[X]$.

We have thus shown that $G^{(1)}[X] - G[X] \cdot (Q \cup U)$ cannot contain a strongly minimal A, i.e., this difference is finite. But U is $\varnothing$-definable, since it is stable under all automorphisms of $\mathfrak{M}$. Hence it follows that the difference is a subset of $G[X]$, but this means that it is empty. □

REMARK 5.3. We have proved incidentally that if $S \subseteq G^{(1)}[\varnothing]$, S is a strongly minimal subgroup and $g \cdot S \cap G^{(1)}[X]$ is infinite, then $g \cdot S = g' \cdot S$ for some $g' \in G[X]$. □

LEMMA 5.4. *Any strongly minimal subgroup of* G *is a subset of* $G^{(1)}[\varnothing]$ *and the number of such subgroups is finite.*

PROOF. It will be convenient in this proof to use additive notation for H. Let K be the set of all strongly minimal subgroups of $H \times H$ other than $\{0\} \times H$. We define multiplication and addition operations on the set K as follows:

First, for arbitrary $S, T \in K$ denote

$$S + T = \{\langle h, s + t\rangle : \langle h, s\rangle \in S,\ \langle h, t\rangle \in T\},$$
$$S \circ T = \{\langle s, t\rangle : (\exists h \in H)(\langle s, h\rangle \in S \,\&\, \langle h, t\rangle \in T)\}.$$

Obviously, $S + T$ and $S \circ T$ are subgroups of rank 1, but they are not necessarily strongly minimal. However, each such subgroup contains a uniquely determined strongly minimal subgroup—its connected component. We consider this subgroup the result of the addition or multiplication operation, as the case may be. It is easy to see that K is a division ring relative to these operations. This division ring is locally finite, since $\mathfrak{M}$ is $\aleph_0$-categorical. By the Wedderburn-Artin Theorem, K is a field.

Let S be an arbitrary subgroup of K defined by a formula $\varphi(\overline{x}, \mathfrak{M})$. If $\overline{y}$ realizes a type $\operatorname{tp}(\overline{x}, \varnothing)$, then the subgroup $T = \varphi(\overline{y}, \mathfrak{M})$, as an element of the multiplicative group of K, is of the same order as S. But since there are only finitely many elements of a given finite order in a field, it follows that there are only finitely many subgroups $\varphi(\overline{y}, \mathfrak{M})$, where $\overline{y}$ realizes $\operatorname{tp}(\overline{x}, \varnothing)$. Hence it follows that the set

$$U' = \bigcup \{\varphi(\overline{y}, \mathfrak{M}) : \operatorname{tp}(\overline{y}, \varnothing) = \operatorname{tp}(\overline{x}, \varnothing)\}$$

is $\varnothing$-definable and has rank 1. Therefore $U' \subseteq G^{(1)}[\varnothing]$, and since $S \subseteq U'$, this completes the proof of the lemma. □

LEMMA 5.5. *For any strongly minimal X-definable subset Q of $G = H \times H$, there exist a strongly minimal subgroup S, $S \subseteq G^{(1)}[\varnothing]$, and an element $g \in G[X]$ such that $Q \sqsubset g \cdot S$.*

PROOF. By 5.3 and 5.4, it will suffice to show that the intersection of Q with some coset of a strongly minimal subgroup of G is infinite. We may assume without loss of generality that $X = \varnothing$ and Q is an atom. In addition, we may assume that $H[\varnothing] = \{1\}$, $G[\varnothing] = \{1\}$. Indeed, this is always the case for the group $\overline{H} = H/H[\varnothing]$, $\overline{H} \times \overline{H} \cong \overline{G} = G/G[\varnothing]$. But if the conclusion of the lemma is true for the image $\overline{Q}$ of Q in $\overline{G}$ then $Q \sqsubset g \cdot S$ for some subgroup S of rank 1, and then Q almost coincides with some coset of a strongly minimal subgroup of S.

Suppose, then, that Q does not satisfy the requirement of the lemma. Then $\mathscr{P}(G, Q)$ is a quasidesign. We notice now that, by Lemma 5.2,

$$G^{(1)}[\varnothing] = Q \cup U.$$

Let $H_i, \dots, H_n$ be all the distinct strongly minimal subgroups of G, $H_1 = \{1\} \times H$, $H_2 = H \times \{1\}$.

For an arbitrary finite subset X we denote

$$x = \|Q[X]\|, \quad \eta_i(x) = \|H_i[X]\|, \qquad i \in \{1, \dots, n\},$$
$$\eta(x) = \|H[X]\|, \qquad \gamma(x) = \|G[X]\|.$$

By Theorem 1.5.5, $\gamma(x)$, $\eta_1(x)$, $\eta(x)$ are polynomials in x with rational coefficients.

Obviously, $\eta_1(x) = \eta_2(x) = \eta(x)$. In addition, for all $i \in \{1, \dots, n\}$,

(1) $\eta_i(x) = \eta(x)$.

To prove this, let H_i be a strongly minimal subgroup of G other than H_1 or H_2. Define

$$B = \bigcup\{S : S \subseteq G^{(1)}[\varnothing],\ S \neq H_1, H_2, S \text{ a strongly minimal subgroup}\}.$$

Obviously, B is $\varnothing$-definable, and therefore $B \cap H_1$ and $B \cap H_2$ are $\varnothing$-definable. In addition, these two sets are finite, and therefore, since $H[\varnothing] = \{1\}$, we obtain $B \cap H_1 = \{1\}$, $B \cap H_2 = \{1\}$. Hence the intersections $H_i \cap H_1$ and $H_i \cap H_2$ are trivial.

Consider the projections $\langle h_1, h_2\rangle \to h_1$ and $\langle h_1, h_2\rangle \to h_2$ of H_i onto H. Obviously, the kernels of these projections are subsets of $H_i \cap H_1$ and $H_i \cap H_2$, respectively, so these kernels are trivial, i.e., H_i is the graph of an automorphism $H \to H$ and $H_i[X]$ the graph of an automorphism $H[X] \to H[X]$, $\|H_i[X]\| = \|H[X]\|$.

Considering the projection of Q onto $H - \{1\}$ and taking into account that Q and $H - \{1\}$ are atoms, one easily shows that

(2) $x = m \cdot (\eta(x) - 1)$ for some integer m.

We now show that for $x' = \|Q[X']\|$, $X' = X \cup \{y\}$, $y \in Q - Q[X]$,

(3) $\gamma(x') - \gamma(x) = \gamma(x)(n + m\eta(x))(\eta(x')/\eta(x) - 1)$, where n is the number of strongly minimal subgroups of G.

Observe, first of all, that since $R(y, X) = 1$,

$$G[X'] \subseteq G^{(1)}[X].$$

Therefore, by Lemma 5.2,

$$G[X'] = G[X] \cdot (Q[X'] \cup U[X']),$$

that is,

$$\gamma(x') = \|G[X] \cdot (Q[X'] \cup U[X'])\|.$$

We now note that since $H_i \cap g \cdot Q$ is finite for any group H_i, it follows that

$$G[X] \cdot Q[X'] \cap G[X] \cdot U[X'] \subseteq G[X],$$

and therefore

(4) $\gamma(x') - \gamma(x) = \|G[X] \cdot Q[X'] - G[X]\| + \|G[X] \cdot U[X'] - G[X]\|$.

Clearly, $G[X] \cdot Q[X'] - G[X] = G[X] \cdot (Q[X'] - Q[X])$. We claim that if $g_1, g_2 \in G[X]$ and $g_1 \neq g_2$, then

$$g_1 \cdot (Q[X'] - Q[X]) \cap g_2 \cdot (Q[X'] - Q[X]) = \varnothing.$$

Indeed, we may assume that $g_2 = 1$. Since $Q[X'] - Q[X] \not\subseteq \operatorname{acl}(g_1)$, the assumption that the intersection is nonempty would imply that

$$g_1 \in H_0 = \{g \in G\colon g \cdot Q \cap Q \text{ is infinite}\}.$$

Since Q satisfies condition (3.2), H_0 is finite, so $H_0 \subseteq G[\varnothing] = \{1\}$.

It follows from what we have proved that

(5) $\|G[X] \cdot Q[X'] - G[X]\| = \gamma(x)(x' - x)$.

Let $i \neq j$, $i, j \in \{1, \dots, n\}$, $g_i, g_j \in G[X]$, $h_i \in H_i$, $h_j \in H_j$. Let $g_i \cdot h_i = g_j \cdot h_j$ or, equivalently, $g_j^{-1} \cdot g_i = h_j \cdot h_i^{-1}$. Since $g_j^{-1} \cdot g_i = g \in G[X]$ and there are only finitely many pairs $\langle h_j, h_i \rangle$ for which $g = h_j \cdot h_i^{-1}$, it follows that $h_j \cdot h_i^{-1} \in G[X]$. Hence it follows that

$$\begin{aligned} \|G[X] \cdot U[X'] - G[X]\| &= \|G[X] \cdot H_1[X'] - G[X]\| \\ &\quad + \cdots + \|G[X] \cdot H_n[X'] - G[X]\|. \end{aligned} \tag{6}$$

Since $G[X]$ and $H_i[X']$ are groups and $G[X] \cap H_i[X'] = H_i[X]$, we have

(7) $\|G[X] \cdot H_i[X']\| = \|G[X]\| \cdot \|H_i[X']\colon H_i[X]\|$.

We now deduce from (4), (5), (6), (7), and (1) that

(8) $\gamma(x') - \gamma(x) = \gamma(x)(x' - x) + \gamma(x) \cdot n \cdot (\frac{\eta(x')}{\eta(x)} - 1)$.

In view of (2), this finally yields (3).

From the fact that $G = H \times H$, it easily follows that for all x

$$\gamma(x) = \eta(x)^2.$$

Hence (3) implies

$$\left(\frac{\eta(x')}{\eta(x)}\right)^2 - 1 = (n + m \cdot \eta(x)) \left(\frac{\eta(x')}{\eta(x)} - 1\right),$$

or, since $\eta(x') > \eta(x)$,

(9) $\frac{\eta(x')}{\eta(x)} + 1 = n + m \cdot \eta(x)$.

We now observe that, since H is a strongly minimal group, it follows that H is a group of a prime exponent p, that is to say,

(10) $\eta(x) = p^z$

for a suitable natural number z. We now obtain from (9)

(11) $p^{z'-z} = n - 1 + m \cdot p^z$.

Since $n > 1$, equality (11) yields a contradiction when $p^z > n - 1$. □

COROLLARY 5.6. *If T is an X-definable subset of rank 1 of the group $G = H \times H$, there exist a strongly minimal subgroup $S \subseteq G^{(1)}[\varnothing]$ and an element $g \in G[X]$ such that $g \cdot S \cap T$ is infinite.*

PROOF. By Shelah's Finite Equivalence Relation Theorem, there exists in T a strongly minimal set Q which is acl(X)-definable. But since $G[\text{acl}(X)] = G[X]$, it follows that g and S as in Lemma 5.5 satisfy the requirements of the corollary. □

We now go on directly to construct the vector space required in Proposition 5.1. Let H be the group defined above, $\overline{H} = H/H[\varnothing]$. As already noted, $\overline{H}[\varnothing] = \{1\}$. The group $\overline{H}$, with additive notation, will be denoted by V.

It follows from 5.4 that there are only finitely many relatively definable strongly minimal subgroups in $V \times V$. As shown in the proof of 5.5, each of these subgroups, with the exception of $\{0\} \times V$ and $V \times \{0\}$, is the graph of an automorphism of V. Let K be the set of all such automorphisms together with the zero homomorphism. By construction, K is the set of all definable homomorphisms of V into V and K is finite. K is closed under addition and superposition of homomorphisms, so K is a field. Moreover, K acts on V, which thereby becomes a vector space over K.

We now extend the signature Ω by constant symbols to Ω', in such a way that all automorphisms in K become $\varnothing$-definable. Since the graphs of all these automorphisms are strongly minimal subgroups of $G^{(1)}[\varnothing]$, $R(G^{(1)}[\varnothing]) = 1$, it follows that each of them almost coincides with a certain strongly minimal acl($\varnothing$)-definable subset of $G^{(1)}[\varnothing]$ (by Shelah's Finite Equivalence Relation Theorem). Hence it follows at once that all automorphisms in K are acl($\varnothing$)-definable. In other words, Ω' may be obtained from Ω by the addition of finitely many constants from acl($\varnothing$) (understood in an extension of $\mathfrak{M}$ by definition).

LEMMA 5.7. *For any $X \subseteq V$, the algebraic closure $V[X]$ of X in V coincides with the linear space of $X \cup V[\varnothing]$.*

PROOF. We may assume that X is finite, $X = \{x_1, \dots, x_n\}$, the x_i being algebraically independent elements. We proceed by induction on n.

For $n = 0$, $X = \varnothing$, the assertion is obvious, since $V[\varnothing]$ is its own linear span.

Let $X' = X - \{x_n\}$, $y \in V[X]$. Then

$$R(\langle x_n, y\rangle, X') = 1 = R(\langle x_n, y\rangle, V[X']),$$

and consequently there exists an X'-definable set T containing $\langle x_n, y\rangle$ such that $R(T) = 1$, $T \subseteq V \times V$. We may assume that T is a $V[X']$-atom. By 5.6, there exist a strongly minimal subgroup $S \subseteq V \times V$ and an element $\langle v, w\rangle$ of $V[X'] \times V[X']$ such that $(\langle v, w\rangle + S) \cap T$ is infinite. S is the graph of some automorphism k in K; in particular, S is $\varnothing$-definable. Since T is a $V[X']$-atom, it follows that $T \subseteq \langle v, w\rangle + S$. Hence it immediately follows that

$$\langle x_n, y\rangle = \langle v, w\rangle + \langle t, kt\rangle \quad \text{or} \quad y = w - kv + kx_n.$$

Since $w, v \in V[X']$, it follows by induction that y is an element of the linear span of X and $V[\varnothing]$. □

PROPOSITION 5.8. *Let V be a vector space over a division ring K, and assume that V is $\varnothing$-definable in an uncountably categorical structure $\mathfrak{M}$; assume that for $X \subseteq V$ $V[X]$ coincides with the linear span of $X \cup V[\varnothing]$. Then any $\varnothing$-definable predicate $S(v_1, \dots, v_n)$ on V is a Boolean combination of predicates of the form*

(*) $k_1 v_1 + \cdots + k_n v_n + c = 0$, $k_1, \dots, k_n \in K$, $c \in V[\varnothing]$.

PROOF BY INDUCTION ON RD(S). Suppose that $R(S) = m$. Then $m \le n$ and there exist $x_1, \dots, x_n$ such that $S(x_1, \dots, x_n)$, and exactly m of the elements $x_1, \dots, x_n$ are algebraically independent. Suppose these are $x_1, \dots, x_m$ and let $x_{m+i} \in \operatorname{acl}(x_1, \dots, x_m)$ for $i = 1, \dots, n-m$. By assumption,

$$x_{m+i} = k_1^i x_1 + \cdots + k_m^i x_m + c^i, \qquad c^i \in V[\varnothing],\ k_1^i, \dots, k_m^i \in K.$$

Let

$$\begin{aligned} P = \{(v_{m+i} = k_1^i v_1 + \cdots + k_m^i v_m + c^i) &: i = 1, \dots, n-m\} \\ \cup\{(k_1 v_1 + \cdots + k_m v_m \ne c) &: c \in V[\varnothing],\ k_1, \dots, k_m \in K, \\ &\qquad k_1 \ne 0 \vee \cdots \vee k_m \ne 0\}. \end{aligned}$$

By assumption, p is the complete type realized by the tuple $\langle x_1, \dots, x_n\rangle$. Therefore,

$$Th(M) \cup p \vdash S(v_1, \dots, v_n), \qquad M \vDash P(v_1, \dots, v_n) \to S(v_1, \dots, v_n)$$

for some finite conjunction P of formulas in p. Since $R(\langle x_1, \dots, x_n\rangle, \varnothing) = m$ and $\langle x_1, \dots, x_n\rangle$ satisfies P, we have $R(P) = m$. Therefore RD$(S \,\&\, \neg P)$

$< \mathrm{RD}(S)$. By the inductive hypothesis, $S \& \neg P$ is equivalent to a Boolean combination of formulas of the form $(*)$. Consequently, this is also true for S, because $S \equiv S \& \neg P \vee P$. □

COROLLARY 5.9. *If A is an affine space, $\varnothing$-definable in $\mathfrak{M}$ and associated with the vector space V of Proposition 5.8, then any $\varnothing$-definable predicate $S(w_0, w_1, \dots, w_n)$ on A is a Boolean combination of predicates of the form*

$$k_1\overrightarrow{w_0, w_1} + \cdots + k_n\overrightarrow{w_0w_n} + c = 0, \qquad k_1, \dots, k_n \in K,$$

where c is a parallel translation of A (an element of $V[\varnothing]$).

For the proof it suffices to observe that the map $\langle w_0, w\rangle \to \overrightarrow{w_0w}$ carries predicates defined on A over to V. □

Proposition 5.8 completes the proof of 5.1, and hence also of Theorem 0.1.

It should be noted that assertions 5.2–5.8 contain the proof of the following

PROPOSITION 5.10. *If V is a strongly minimal group, $\varnothing$-definable in a totally categorical structure $\mathfrak{M}$, and $V[\varnothing] = \{0\}$, then one can endow V with the structure of a vector space* $\mathrm{acl}(\varnothing)$*-definable in $\mathfrak{M}$, in such a way that every $\varnothing$-definable predicate on V is a Boolean combination of predicates of the form $(*)$.*

Combined with the Classification Theorem, this implies

THEOREM 5.11. *Let A be a strongly minimal countably categorical structure such that the algebraic closure of every point of A contains that point alone. Then only one of the following cases is possible*:

(a) *The geometry A is degenerate; A is an indiscernible set.*

(b) *The geometry A is affine; A is the set of points of some vector space V over a finite field K with a distinguished subspace of constants V_0, and every $\varnothing$-definable predicate on A is a combination of predicates of the form*

$$k_1(v_1 - v_0) + \cdots + k_n(v_n - v_0) + c = 0, \qquad k_1, \dots, k_n \in K, \ c \in V_0.$$

(c) *The geometry A is projective. There exist a vector space V over a finite field K and a canonical projection of the pregeometry of V onto the geometry of the projective space A, such that any $\varnothing$-definable predicate on A is a Boolean combination of projections onto A of predicates on V of the form*

$$k_1v_1 + \cdots + k_nv_n = 0, \qquad k_1, \dots, k_n \in K.$$

PROOF. The partition into three cases follows from Theorem 0.1. If the geometry A is degenerate, then (a) is obvious. If A is affine, (b) follows from 1.1, 5.8, 5.9.

If A is a projective geometry and V a vector space whose projection is A, consider the structure $\mathfrak{N}$ obtained by joining to A the space V together with the graph of the projection of V onto A. It is easy to verify that $\mathfrak{N}$

is a totally categorical structure and that predicates which are $\varnothing$-definable on A in the sense of the structures $\mathfrak{A}$ and $\mathfrak{N}$ coincide. In addition, $V[X]$ is the preimage of $A[X]$ for any X in $\mathfrak{N}$. Since $A[\varnothing] = \varnothing$, it follows that $V[\varnothing] = \{0\}$. Since $A[X]$ is the projective subspace spanned by X for $X \subseteq A$, it follows that $V[X]$ is the linear span of X for $X \subseteq V$. Part (c) now follows immediately from 5.8. $\square$

CHAPTER IV

Description of Strongly Minimal Quasialgebras

The present chapter is devoted to an important class of strongly minimal structures, called here quasialgebras.

A structure $\mathfrak{A}$ is called a *quasialgebra* if, for any $X \subseteq A$, $A[X] = A \cap \mathrm{dcl}(X)$.

We recall that here $\mathrm{dcl}(X)$ is the union of all X-definable singletons. Generally speaking, we shall consider the case in which $\mathfrak{A}$ is a structure which is $\varnothing$-definable in some structure $\mathfrak{M}$, in which case $\mathrm{acl}(X)$ and $\mathrm{dcl}(X)$ are understood in the sense of $\mathfrak{M}$.

Here are three typical examples of quasialgebras:

(1) $\langle A, \{g: g \in G\}\rangle$, where A is a set, G a group acting on A, every element g of G other than the identity defines a bijection of A onto itself, whose fixed points are precisely the points of a fixed subset $A_0 \subseteq A$.

(2) A vector space over a division ring with a subspace of constants.

(3) An affine space over a division ring with a subspace of distinguished parallel translations.

An important class of quasialgebras was introduced by Marczewski and Narkiewicz [65] and described by Urbanik [75].

A set A with operations is called a v^**-algebra* if the condition: for some $a_1, \dots, a_n \in A$ and terms f, g such that $f(v_1, \dots, v_n) \neq g(v_1, \dots, v_n)$ it is true that $f(a_1, \dots, a_n) = g(a_1, \dots, a_n)$ implies that

$$a_{i+1} = h(a_1, \dots, a_i, a_{i+2}, \dots, a_n)$$

or some $i \in \{0, \dots, n-1\}$ and a suitable term h.

It is easily deduced directly from the definition that an infinite v^*-algebra is a strongly minimal quasialgebra, with the exception of the case of infinite A_0 in example (1).

Urbanik described all v^*-algebras of dimension 3 and higher as the algebras listed in examples (1)–(3).

Finally, the most interesting class of strongly minimal quasialgebras from a model-theoretical standpoint we find to be the uncountably categorical universally axiomatizable structures. Indeed, let $\mathfrak{M}$ be such a structure. By a well-known theorem of Lindström [55] the theory of $\mathfrak{M}$ is model complete, i.e., any infinite substructure of $\mathfrak{M}$ is an elementary submodel.

In particular, if $X \subseteq M$, $\mathrm{fcl}(X)$ is the closure of X relative to all the operations and $\mathrm{fcl}(X)$ is infinite, then $\mathrm{fcl}(X)$ is an elementary submodel of $\mathfrak{M}$. Clearly, $\mathrm{fcl}(X) \subseteq \mathrm{dcl}(X)$. On the other hand, $\mathrm{fcl}(X) \supseteq \mathrm{acl}(X)$, since $\mathrm{fcl}(X)$ is a model of the theory. Hence $\mathrm{acl}(X) = \mathrm{dcl}(X)$. Finally, let A be a strongly minimal $\varnothing$-definable subset of M. Then for infinite X $A[X] = A \cap \mathrm{dcl}(X)$. Suppose that this condition fails to hold for finite $X \supseteq A$. Let $\mathrm{acl}(X) \supsetneqq \mathrm{dcl}(X) \supseteq \mathrm{fcl}(X)$, $y \in A[X] - \mathrm{dcl}(X)$. Since $\mathrm{acl}(X') = \mathrm{fcl}(X')$ for any infinite extension $X' \supseteq X$, it follows that $\mathrm{acl}(X) \subseteq \mathrm{fcl}(X')$, that is, $y \in \mathrm{fcl}(X')$. We may assume that $y \in \mathrm{fcl}(X \cup \{z\})$ for $z \in A - A[X \cup \{y\}]$; more precisely: $y = f(x_1, \dots, x_n, z)$, where f is an $(n+1)$-place term, $x_1, \dots, x_n \in X$. Since $z \notin \mathrm{acl}(X \cup \{y\})$, it follows that $Z = \{z' \in A: f(x_1, \dots, x_n, z') = y\}$ is infinite. But this means that $A - Z$ is finite, consisting, say of m elements. It is now easy to see that y is the unique element satisfying the following condition for v:

$$(\exists_w^{>m})(w \in A \;\&\; v = f(x_1, \dots, x_n, w)).$$

This proves that $y \in \mathrm{dcl}(X)$, that is, $A[X] = A \cap \mathrm{dcl}(X)$ for all $X \subseteq A$, and A is a strongly minimal quasialgebra.

In the case that the theory of $\mathfrak{M}$ defines a quasivariety, it is not hard to define a v^*-algebra in $\mathfrak{M}$ [24]. This fact has been used to describe the uncountably categorical varieties using Urbanik's theorem.

The main theorem of this chapter gives a description of an arbitrary strongly minimal quasialgebra. It turns out that all such quasialgebras reduce, up to definability, to examples (1)–(3) (for the precise formulation see §3). One thus has a description, up to definability, of the strongly minimal subsets A of models $\mathfrak{M}$ of universally axiomatizable uncountably categorical theories. This also largely characterizes the model as a whole, since $M \subseteq \mathrm{fcl}(A)$ in this case; moreover, M is definable in the structure $\mathfrak{A}$. Indeed, since for any x in M there are a term f and elements $y_1, \dots, y_n \in A$ such that $x = f(y_1, \dots, y_n)$, it is easy to show by induction on $\mathrm{RD}(M)$ that there exists a $\varnothing$-definable map g of A^n onto M. This implies that M is definable in A.

Finally, our proof of the theorem on strongly minimal quasialgebras is a new proof of Urbanik's theorem for the case of infinite v^*-algebras.

The idea of the proof of the main theorem in this chapter is based on the proof of the main theorem of Chapter 3. In particular, the fundamental fact, Lemma 2.6, is an analog of Proposition 3.1.4, and in this chapter as in Chapter 3 the translation group on A plays an essential role.

§1. Translations on a strongly minimal quasialgebra

Throughout this chapter A will be a strongly minimal quasialgebra. We shall use the definitions and some facts about quasitranslations on A from Chapter 3.

Let $\mathscr{O}^n$ denote the set of all $\varnothing$-definable partial n-place operations on A, i.e., partial maps $A^n \to A$. Denote $\mathscr{O} = \bigcup_n \mathscr{O}^n$. If $f(\overline{w}, v)$ is an operation in $\mathscr{O}^n$, $\overline{w}$ a tuple of variables of length $n-1$, $\overline{x} = \langle x_1, \dots, x_{n-1}\rangle$, then $f_{\overline{x}}$ will denote the map $v \to f(\overline{x}, v)$.

LEMMA 1.1. (a) *Let*

$$X \subseteq A, \quad z \in A - A[X], \quad y \in A[X \cup \{z\}] - A[X].$$

Then there exist $q \in \mathscr{O}$ *and a quasitranslation* g *of the form* $q_{\overline{x}}$ *such that* $g(z) = y$, $\overline{x}$ *being a tuple of elements of* X.

(b) *Any* X*-definable quasitranslation* g *maps* $A - A[X]$ *bijectively onto itself. Moreover, there exists an* X*-definable quasitranslation* h *such that* $h \circ g$ *is a quasitranslation defined on* A *and equal there to the identity everywhere except at a finite number of points contained in* $A[X]$.

PROOF. (a) By the definition, $y \in \mathrm{dcl}(\{z\} \cup X)$. This means that for some formula $\varphi \in F_{n+2}$ and elements $x_1, \dots, x_n \in X$

$$\mathfrak{M} \models \varphi(\overline{x}, z, y) \ \&\ (\exists! w)\varphi(\overline{x}, z, w),$$

where $\overline{x} = \langle x_1, \dots, x_n\rangle$. We can adjust φ so that, in addition,

$$\mathfrak{M} \models (\forall \overline{v})((\exists w)(\varphi(\overline{v}, w)) \to (\exists! w)\varphi(\overline{v}, w)),$$

where $\overline{v} = \langle v_1, \dots, v_n, v_{n+1}\rangle$; φ gives the graph of the required map.

(b) If $g(z) = y$, $y \in A[X \cup \{z\}] - A[X]$, then $z \in A[X \cup \{y\}]$. Consequently, by what we have already proved, there exists an X-definable quasitranslation h such that $h(y) = z$, that is, $h \circ g(z) = z$. Since $h \circ g$ is X-definable and $z \notin A[X]$, it follows that the set of points where $h \circ g$ is defined and identical is an infinite X-definable subset of A.

REMARK. Lemma 1.1 in our case replaces Lemma 3.1 of the previous chapter and enables us to consider the translation group $T = QT/E$ on A.

An X-definable quasitranslation g is said to be *linear* if there exists $q \in \mathscr{O}^3$ such that for any $a \in A - A[X]$ there exists $b \in A$ for which the quasitranslation q_{ab} coincides with g modulo E.

Otherwise the quasitranslation is said to be nonlinear.

REMARK. If gEq_b for $q \in \mathscr{O}^2$, $b \in A$, then g is obviously linear.

REMARK. It is easy to see that linearity of a quasitranslation is preserved by the congruence E; it is therefore legitimate to speak of linearity of translations.

LEMMA 1.2. *An* X*-definable quasitranslation* g *is linear if and only if, for some pair* c_1, c_2 *of elements of* A *which are algebraically independent over* X,

$$g(c_2) \in \mathrm{acl}(c_1, g(c_1), c_2).$$

PROOF. *Necessity*. If g is a trivial quasitranslation, then $g(c) \in \mathrm{acl}(c)$ for $c \in A - A[X]$. We may therefore assume that g is nontrivial. By the definition of linearity, for $a \in A - A[X]$ there exist $q \in \mathscr{O}^3$ and $b \in A$ such that

$$g(c_1) = q_{ab}(c_1) \in \operatorname{acl}(a, b, c_1), \qquad g(c_2) = q_{ab}(c_2) \in \operatorname{acl}(a, b, c_2)$$

for any $c_1, c_2 \in A - A[X \cup \{a, b\}]$. Choose c_1, c_2 independent over $X \cup \{a, b\}$. Since q_{ab} is nontrivial, either

$$a \in \operatorname{acl}(g(c_1), b, c_1) \quad \text{or} \quad b \in \operatorname{acl}(g(c_1), a, c_1),$$

say the latter. Then

$$g(c_2) \in \operatorname{acl}(a, g(c_1), c_1, c_2).$$

Hence,

$$g(c_2) \in \operatorname{acl}(c_1, g(c_1), c_2),$$

since otherwise

$$a \in \operatorname{acl}(g(c_1), g(c_2), c_1, c_2) \subseteq \operatorname{acl}(X \cup \{c_1, c_2\}),$$

contrary to the choice of c_1, c_2.

Sufficiency. If

$$g(c_2) \in \operatorname{acl}(c_1, g(c_1), c_2),$$

then by Lemma 1.1(a) there exists a quasitranslation $q_{\overline{x}}$ which coincides with g modulo E, where $\overline{x}$ is a tuple of elements from $\{c_1, g(c_1)\}$. Since the latter set consists of two elements, we may assume that $q \in \mathscr{O}^3$, $\overline{x} = \langle c_1, g(c_1)\rangle$. Since c_1 is algebraically independent of c_2 over X it follows, taking an arbitrary $a \in A - A[X]$, $b = g(a)$, that gEq_{ab}, and this means that g is linear. □

REMARK 1.3. The above argument shows, in particular, that one can take $b = g(a)$ in the definition of linearity, and consequently $b \in A[X \cup \{a\}]$.

PROPOSITION 1.4. *If all quasitranslations on A are trivial, then one can define an action of the translation group T on $A - A[\varnothing]$ so that for every $t \in T$, other than the identity, the map $a \to ta$ is a $\varnothing$-definable bijection of $A - A[\varnothing]$ without fixed points. For any $X \subseteq A - A[\varnothing]$,*

$$A[X] - A[\varnothing] = \{tx \colon t \in T, x \in X - A[\varnothing]\}.$$

Every $\varnothing$-definable n-place relation $Q(v_1, \dots, v_n)$ restricted to $A - A[\varnothing]$ is a Boolean combination of relations of the type $v_i = tv_j$, where $i, j \in \{1, \dots, n\}$, $t \in T$.

PROOF. Since any quasitranslation g on A is trivial, we have $g(a) \in \operatorname{acl}(a)$ for all $a \in A - A[X]$. Therefore, by Lemma 1.1, gEg', where g' is a $\varnothing$-definable quasitranslation defining the action of the translation $t = gE$ on $a \in A - A[\varnothing]$: $ta = g'(a)$.

By the same Lemma 1.1(a), if $y \in A[X] - A[\varnothing]$, then $y = q_{x_1, \dots, x_n}(x)$, where $x_1, \dots, x_n, x \in X$, $x \notin \operatorname{acl}(x_1, \dots, x_n)$, $q_{x_1, \dots, x_n}$ is a quasitranslation. Since this quasitranslation is trivial, it follows that $y = g(x)$ for some $\varnothing$-definable g, that is, $y = tx$, where t is the translation gE.

Finally, the last assertion is proved by induction on $\mathrm{RD}(Q)$. If $R(Q) = m$, then $m \le n$ and there exist $a_1, \dots, a_n \in A - A[\varnothing]$ such that $Q(a_1, \dots, a_n)$ and m of the elements $a_1, \dots, a_n$ are algebraically independent, say these are $a_1, \dots, a_m$, and moreover $a_{m+1}, \dots, a_n \in \mathrm{acl}(a_1, \dots, a_m)$. This implies that for every $j \in \{m+1, \dots, n\}$ there exist $t_j \in T$ and $a_{i(j)}$, $1 \le i(j) \le m$, such that $a_j = t_j a_{i(j)}$. Let

$$p(v_1, \dots, v_n) = \{v_i \ne t v_k : 1 \le i < k \le m, t \in T\} \\ \cup \{v_i \ne c : 1 \le i \le m, c \in A[\varnothing]\} \cup \{v_j = t_j v_{i(j)} : m < j \le n\}.$$

It follows from the description of the algebraic closure obtained above that p is the complete type realized by the tuple $\langle a_1, \dots, a_n \rangle$. Consequently,

$$Th(\mathfrak{A}) \cup p(v_1, \dots, v_n) \vdash Q(v_1, \dots, v_n),$$

and so for some formula P & P_1

$$Th(\mathfrak{A}), P_0(v_1, \dots, v_n) \,\&\, P_1(v_1, \dots, v_n) \vdash Q(v_1, \dots, v_n),$$

where P_0 is a finite conjunction of formulas of the form $v_i = t v_j$ or $v_i \ne t v_j$ in p, and P_1 is a finite conjunction of formulas of the form $v_i \ne c$ ($c \in A[\varnothing]$) in p.

Obviously, $R(P_0 \,\&\, P_1) = m = R(Q)$, and therefore

$$\mathrm{RD}(Q \,\&\, \neg(P_0 \,\&\, P_1)) < \mathrm{RD}(Q).$$

By the inductive hypothesis, the relation $Q \,\&\, \neg(P_0 \,\&\, P_1)$ on $A - A[\varnothing]$ is a Boolean combination Q' of the required form. On the other hand, $P_0 \,\&\, P_1$ is obviously equivalent on $A - A[\varnothing]$ to P_0, and therefore Q is equivalent on $A - A[\varnothing]$ to $Q' \vee P_0$—a Boolean combination of formulas of the form $v_i = t v_j$, $t \in T$. □

§2. Nonlinear translations

The aim of this section is to prove that any translation defined on a strongly minimal quasialgebra is linear. The proof will be done by *reductio ad absurdum,* assuming the existence of a nonlinear translation and investigating the translation group T in that case.

By adding an arbitrary set of constants X to the signature, we obviously again get a strongly minimal quasialgebra A_X from the strongly minimal quasialgebra A.

LEMMA 2.1. *Let* $q \in \mathscr{O}^{n+1}$, $n \ge 3$, $\overline{x} = \langle x_1, \dots, x_n \rangle \in A^n$, *where* $x_1, \dots, x_n$ *are algebraically independent, and let* $q_{\overline{x}}$ *be a nonlinear translation on* A. *Then for some* $i \in \{1, 2, 3\}$ $q_{\overline{x}}$ *is a nonlinear quasitranslation on* A_{x_i}.

PROOF. Suppose the contrary; then by Lemma 1.2 there are elements c_1, c_2 in A, algebraically independent over X, such that for all $i \in \{1, 2, 3\}$,

$$q_{\overline{x}}(c_2) \in \mathrm{acl}(c_1, q_{\overline{x}}(c_1), c_2, x_i) - \mathrm{acl}(c_1, q_{\overline{x}}(c_1), c_2);$$

therefore

$$x_1, x_2, x_3 \in \operatorname{acl}(c_1, q_{\overline{x}}(c_1), c_2, q_{\overline{x}}(c_2)),$$

and hence

$$\dim A[c_1, c_2, x_1, x_2, x_3] \leq 4,$$

contrary to the choice of c_1, c_2. □

COROLLARY. *If there exists a nonlinear quasitranslation on* A, *then in some enrichment* A_X *there exists a* $\{y_1, y_2\}$*-definable nonlinear quasitranslation.*

Thus, enriching the signature of the quasialgebra, if necessary, by a finite set of constants, we may assume that there exists a nonlinear quasitranslation on A of the form $q_{y_1y_2}$, where $q \in \mathcal{O}^3$. Throughout this section, the letter q will be reserved for this kind of operation on $\mathcal{O}^3$.

LEMMA 2.2. (a) $q_{y_1y_2}$ *is a nonlinear quasitranslation if and only if* y_1, y_2 *are algebraically independent elements.*

(b) *There exists* $q^* \in \mathcal{O}^3$ *such that* $q_{y_1y_2}(z_1) = z_2 \Leftrightarrow q^*_{z_1z_2}(y_1) = y_2$ *for* $z_1, z_2 \in A - A[y_1, y_2]$.

(c) *If*

$$q_{y_1y_2} E q_{x_1x_2}$$

for some $x_1, x_2 \in A$, *then* $\operatorname{acl}(x_1, x_2) = \operatorname{acl}(y_1, y_2)$. *In other words, there exist only finitely many pairs* $\langle x_1, x_2 \rangle$ *for which* $q_{x_1x_2}$ *coincides with* $q_{y_1y_2}$ *modulo* E.

PROOF. (a) Let $q_{y_1y_2}$ be nonlinear, $z_1 \in A - A[y_1, y_2]$, $z_2 = q_{y_1y_2}(z_1)$. Then $y_1 \notin A[y_2]$, for otherwise $z_2 \in s_{y_2}(z_1)$ for a suitable $s \in \mathcal{O}^2$, contrary to the nonlinearity of $q_{y_1y_2}$.

If $q_{y_1y_2}$ is nonlinear for some pair of algebraically independent elements, it will also be nonlinear for any such pair y_1, y_2.

(b) It follows from the definition of linearity that for the stipulated choice of z_1, z_2

$$y_2 \in A[z_1, z_2, y_1],$$

which means

$$y_2 = q^*(z_1, z_2, y_1),$$

for some $q^* \in \mathcal{O}^3$. We now observe that by Lemma 1.1(b) the map $v \to q(y_1, v, z_1)$ is injective, and therefore

$$q(y_1, y_2, z_1) = z_2 \Leftrightarrow q^*(z_1, z_2, y_1) = y_2$$

for any algebraically independent y_1, y_2, z_1.

(c) It follows at once from the definition of linearity that $x_1, x_2 \in A[y_1, y_2]$. By (a), x_1, x_2 are algebraically independent, and so $A[x_1, x_2] = A[y_1, y_2]$. □

LEMMA 2.3. *Let* g *be an* X*-definable quasitranslation on* A, a_1, a_2 *algebraically independent elements over* X, y_1, y_2 *algebraically independent elements of* A, $a_1, a_2 \notin A[y_1, y_2]$,

$$q_{y_1y_2}(a_1) = g(a_1), \qquad q_{y_1y_2}(a_2) = g(a_2).$$

Then $y_1, y_2 \in A[X]$ *and* $q_{y_1y_2}Eg$.

PROOF. We first observe that

$$q^*(a_1, g(a_1), y_1) = y_2 = q^*(a_2, g(a_2), y_1).$$

Consider the map h defined by

$$h(v) = q^*(v, g(v), y_1),$$

which is obviously $X \cup \{y_1\}$-definable. We note that $y_2 \in A[X \cup \{y_1\}]$, for otherwise h would contradict Lemma 1.1. Since $A[X \cup \{y_1\}] \neq A[X \cup \{a_1, a_2\}]$, by virtue of the different dimensions of these sets over X, it follows that $a_1 \notin A[X \cup \{y_1\}]$ or $a_2 \notin A[X \cup \{y_1\}]$. Let us assume that the second of these assertions is true. Then

$$q_{y_1y_2}(a_2) = g(a_2),$$

and this implies $q_{y_1y_2}Eg$, since both quasitranslations are $X \cup \{y_1, y_2\}$-definable, whereas $a_2 \notin A[X \cup \{y_1, y_2\}]$. □

LEMMA 2.4. *For any algebraically independent* a_1, a_2, b_1, b_2 *in* A, *there exist algebraically independent* $y_1, y_2 \in A$ *such that*

$$q_{y_1y_2}(a_1) = b_1, \qquad q_{y_1y_2}(a_2) = b_2.$$

PROOF. If $q_{y_1y_2}$ is nonlinear, then by Lemma 1.2 $q_{y_1y_2}(a_1)$, $q_{y_1y_2}(a_2)$, a_1, a_2 are algebraically independent if that is true of y_1, y_2, a_1, a_2. By homogeneity, we may assume that $b_1 = q_{y_1y_2}(a_1)$, $b_2 = q_{y_1y_2}(a_2)$. □

Lemmas 1.2, 2.3, and 2.4 imply the following result as a corollary.

LEMMA 2.5. *For any nonlinear quasitranslation* g, *there exists a translation* $q_{y_1y_2}$ *equal to it modulo* E.

From now on we shall be working in the translation group, using the notation $s_{\overline{x}}$, where $s \in \mathcal{O}^{n+1}$, $\overline{x} \in A^n$, also for translations, to designate the translation represented by the quasitranslation $s_{\overline{x}}$.

LEMMA 2.6. *Every translation is representable as* $s_{a_1a_2}$ *for some* $a_1, a_2 \in A$, $s \in \mathcal{O}^3$. *Nonlinear translations are representable as* $q_{a_1a_2}$ *for algebraically independent* $a_1, a_2 \in A$.

Proof. The lemma follows from 2.5 and the definition of linearity. □

Lemma 2.7. *For any translation t, there exist nonlinear translations $q_{a_1a_2}$ and $q_{b_1b_2}$ such that $t = q_{a_1a_2} \circ q^{-1}_{b_1b_2}$.*

Proof. Let $t = s_{d_1d_2}$, $s \in \mathcal{O}^3$. Choose $b_1, b_2 \in A$ algebraically independent over d_1, d_2 and consider the translation $t' = s_{d_1d_2} \circ q_{b_1b_2}$. We claim that t' is nonlinear, so that we can then put $t' = q_{a_1a_2}$. Suppose the contrary. Then for any $e \in A - A[b_1, b_2, d_1, d_2]$ there exists $c \in A$ such that $t' = r_{ec}$, $r \in \mathcal{O}^3$. Obviously also $q_{b_1b_2} = s^{-1}_{d_1d_2} \circ r_{ec}$, and so, in view of Lemma 2.2(c), $b_1, b_2 \in A[d_1, d_2, e, c]$. Hence $e, c \in A[d_1, d_2, b_1, b_2]$. This contradicts the arbitrariness of the choice of e. □

Proposition 2.8. *If there exists a nonlinear translation, then the group T of all translations is $\varnothing$-definable and connected, $R(T) = 2$.*

Proof. It follows from the preceding lemma that an arbitrary $t \in T$ may be interpreted as a 4-tuple $\langle a_1, a_2, b_1, b_2\rangle \in A^4$ such that $q_{a_1a_2}$, $q^{-1}_{b_1b_2}$ are defined and $t = q_{a_1a_2} \circ q^{-1}_{b_1b_2}$, and such that the following conditions are $\varnothing$-definable:

$$q_{a_1a_2} \circ q^{-1}_{b_1b_2} = q_{a_1'a_2'} \circ q^{-1}_{b_1'b_2'}, \qquad (q_{a_1a_2} \circ q^{-1}_{b_1b_2}) \circ (q_{a_1'a_2'} \circ q^{-1}_{b_1'b_2'}) = q_{a_1''a_2''} \circ q^{-1}_{b_1''b_2''},$$

which means that T is $\varnothing$-definable. By Lemma 2.2(c) $R(q_{a_1a_2}, \varnothing) = 2$, that is, $R(T) \geq 2$. On the other hand, it follows from Lemma 2.6 that for every $t \in T$ there exist $a_1, a_2 \in A$ such that $t \in \operatorname{acl}(a_1, a_2)$, whence $R(T) \leq 2$. Moreover, if $R(t, \varnothing) = 2$, then a_1, a_2 are algebraically independent, $\operatorname{acl}(t) = \operatorname{acl}(a_1, a_2)$ and t is nonlinear, i.e., $t = q_{a_1'a_2'}$ for algebraically independent $a_1', a_2' \in A[a_1, a_2]$. Hence it follows that $D(T) = 1$, i.e., T is a connected group. □

Corollary. *By Cherlin's Theorem (Chapter 1, §4), T is a solvable group. If T is not nilpotent, then the commutator subgroup U of T is a strongly minimal $\varnothing$-definable abelian group isomorphic to the additive group K_+ of some algebraically closed field K; the action of the elements $t \in T$ on U by conjugation is equivalent to the action on K_+ by multiplication by elements of K. The group T is nilpotent if and only if the center of T is infinite.*

We shall show that under our conditions such a description of T leads to a contradiction.

Lemma 2.9. *If T is not nilpotent, then on some definable set $\overline{A}$ such that $\overline{A} \dot{-} A$ is finite one can define operations $+$, $\cdot$ which endow $\overline{A}$ with a structure isomorphic to that of the field K.*

Proof. The commutator subgroup U of T is a strongly minimal $\varnothing$-definable group. If this group contained some nonlinear translation $q_{a_1a_2}$, then, by homogeneity, every nonlinear translation would belong to U, and then, by Lemma 2.7, $U = T$, which is impossible. Thus every translation in U is linear.

Let $t = s_{a_1 a_2}$ be a linear translation in U, $s \in \mathscr{O}^3$, $R(t, \varnothing) = 1$. By assumption, for any $a \in A - A[\varnothing]$ there exists $b \in A$ such that $t = s_{ab}$.

Consider the map

$$\tau_a \colon A \to T, \qquad \tau_a(x) = s_{ax};$$

τ_a is an a-definable map, $t \in U \cap \operatorname{Range} \tau_a$. Therefore, if there existed $a \in A - A[\varnothing]$ for which $U \cap \operatorname{Range} \tau_a$ were finite, it would follow that $t \in \operatorname{acl}(a)$ for every $a \in A - A[\varnothing]$, so that $t \in \operatorname{acl}(\varnothing)$, contrary to the choice of t. Thus $U \cap \operatorname{Range} \tau_a$ is infinite, and this means, in particular, that if $a, x, y \in A$ are algebraically independent, then $s(a, x, y) \notin A[a, y]$, and therefore, by Lemma 1.1, the maps $x \to s(a, x, y)$ and, consequently, also τ_a are injective. Using the fact that A and U are strongly minimal, we obtain: τ_a is a definable bijection of a cofinite subset $A_0 \subseteq A$ onto a cofinite subset of U. Adding to A_0 a finite set of imaginary points, we obtain an a-definable set $\overline{A}$ and a bijection $\overline{\tau}_a$ mapping $\overline{A}$ onto U.

The bijection $\overline{\tau}_a$ carries the operations $+, \cdot$ of the field K onto $\overline{A}$. □

LEMMA 2.10. *The group T is nilpotent.*

PROOF. Otherwise, we have a field structure definable on $\overline{A}$. Let n, $n > 1$, be a natural number relatively prime to the characteristic of the field $\overline{A}$. Then the map $g(v) = v^n$, which is X-definable in $\overline{A}$ for some X, determines a quasitranslation on A. But g is not a bijection of $A - A[X]$, since $y \in A - A[X]$ if and only if $y^n \in A - A[X]$ and the equation $y^n = b$ has n distinct solutions for $b \in A - A[X]$. This contradicts Lemma 1.1(b). □

THEOREM 2.11. *All translations on a strongly minimal quasialgebra are linear.*

PROOF. Our task is to derive a contradiction from the assumption made above that there exists a nonlinear translation.

Let a, b be algebraically independent elements of A. By Lemma 2.4, the set

$$T_a^b = \{q_{y_1 y_2} : q_{y_1 y_2}(a) = b\}$$

is infinite.

Let Z be the center of T, which, by 2.10, is infinite. By Lemma 3.3.2, the partial action of T on A, where tx is defined if $x \notin \operatorname{acl}(t)$, is $\varnothing$-definable. Consider the set

$$Za = \{za \colon z \in Z,\ za \text{ is defined}\}.$$

This set is a-definable; if it is finite, then $za \in A[a]$ for $z \in Z$, that is, z is given by a trivial quasitranslation, contrary to the assumption $R(Z) \geq 1$. Thus Za is infinite, i.e., Za almost coincides with A.

Let $t_1 \neq t_2$, $t_1, t_2 \in T_a^b$, $z \in Z - \mathrm{acl}(t_1) \cup \mathrm{acl}(t_2)$, where t_1, t_2, z are chosen so that $a \notin A[t_1, z] \cup A[t_2, z]$. Then

$$za = x \in A-(A[t_1] \cup A[t_2]), \qquad t_1 x = t_1 za = zt_1 a = zb = zt_2 a = t_2 za = t_2 x.$$

Comparing the first and last terms of this chain of equalities, we obtain a contradiction to the choice of t_1, t_2. □

§3. Vector space over a strongly minimal quasialgebra

As shown in the previous section, all translations on a strongly minimal algebra are linear. By definition, every such translation may be expressed as p_{ab}, where $p \in \mathscr{O}^3$, $a, b \in A$, and for every $a' \in A - A[\varnothing]$ there exists $b' \in A[a, b, a']$ ($b' = p_{ab}(a')$, see Remark 1.3) such that $p_{ab} = p_{a'b'}$. We may assume that this condition holds for all $p \in \mathscr{O}^3$.

LEMMA 3.1. *If a translation t is nontrivial, then $t = p_{ab}$, $p \in \mathscr{O}^3$, where a, b are algebraically independent elements of A. In that case the map $x \to p_{ax}$ of A into T is injective on $A - A[a]$.*

PROOF. If for every representation of t as p_{ab} b is an element of $A[a]$, then for $y \in A - A[a]$ we have $p_{ab}(y) \in A[a, y]$. Since $p_{ab}(y) = p_{a'b'}(y)$ if $a' \in A - A[a, y]$ and b' is a suitable element of $A[a']$, it follows that $p_{ab}(y) = p_{a'b'}(y) \in A[a, y] \cap A[a', y] = A[y]$. Hence p_{ab} is a trivial quasitranslation, contrary to assumption. Thus, for some algebraically independent a, b and some $p \in \mathscr{O}^3$, we have $t = p_{ab}$.

If $y \in A - A[a, b]$, then if $x \in A - A[a, y]$ $p(a, x, y)$ is not an element of $A[y]$. But if $p(a, x, y) \in A[a, y]$, then, since for $a' \in A - A[a, y]$ there exists $x' \in A[a, a', x]$ such that $p_{ax} = p_{a'x'}$ we obtain

$$p_{ax}(y) = p_{a'x'}(y) \in A[a, y] \cap A[a', y] = A[y],$$

which is impossible. Therefore, by Lemma 1.1, the map $x \to p(a, x, y)$ is a quasitranslation bijective on $A - A[a, y]$, whence the assertion of the lemma follows. □

REMARK. In other words, a nontrivial translation t may be expressed as p_{ab} in such a way that once a is fixed, the element b of A uniquely determines t. In particular, if p_{ab}, r_{ac}, s_{ad} are such representations and $p_{ab} \circ r_{ac} = s_{ad}$, then $d \in A[a, b, c]$, $b \in A[a, c, d]$.

LEMMA 3.2. *If $p, r \in \mathscr{O}^3$, e, a, b are algebraically independent elements of A, and p_{ea} is a nontrivial translation, then p_{ea}^{-1} and $p_{ea} \circ r_{eb}$ are nontrivial translations.*

PROOF. The translation $t = p_{ea} \circ r_{eb}$ is linear. If t were trivial, it would follow, taking $x \in A - A[e, a, b]$, that

$$y = p_{ea}(x) = t \circ r_{eb}^{-1}(x),$$

whence $y \in A[e, a, x] \cap A[e, b, x] = A[e, x]$. But for any $e' \in A - A[\varnothing]$ there exist a' and b' such that $p_{ea} = p_{e'a'}$, $r_{eb} = r_{e'b'}$ and this implies that $y \in A[e', x]$ for any $e' \in A - A[y, x]$. Hence $y \in A[x]$, which would mean that p_{ea} were trivial, contrary to assumption.

Similarly one proves that p_{ea}^{-1} is also nontrivial. □

LEMMA 3.3. *If $p, r \in \mathscr{O}^3$, e, a, b are algebraically independent elements, and p_{ea}, r_{ea} are nontrivial translations, then there exist $s \in \mathscr{O}^3$ and $c, d \in A - A[e]$ such that*

$$p_{ea}^{-1} \circ p_{eb} = s_{ec}, \qquad r_{ea}^{-1} \circ r_{eb} = s_{ed}$$

and s_{ec}, s_{ed} are nontrivial translations.

PROOF. Let $a' \in A - A[e, a, b]$,

$$p_{ea} \circ r_{ea'}^{-1} = u_{ex}, \qquad x \in A, u \in \mathscr{O}^3.$$

By the previous lemmas $x \in A[e, a, a'] - (A[e, a] \cup A[e, a'])$, so there exists $b' \in A$ such that

$$\operatorname{tp}(\langle a', a, x, e\rangle, \varnothing) = \operatorname{tp}(\langle b', b, x, e\rangle, \varnothing)$$

and hence

$$p_{eb} \circ r_{eb'}^{-1} = u_{ex},$$

i.e.,

$$p_{ea}^{-1} \circ p_{eb} = r_{ea'}^{-1} \circ r_{eb'} = s_{ec}.$$

Since $a' \in A[e, a, x] - A[e, x]$, $b' \in A[e, b, x] - A[e, x]$, it follows that $A[a', e] \cap A[b', e] = A[e]$, which implies that a', b', e are algebraically independent. Hence $\operatorname{tp}(\langle a', b'\rangle, \{e\}) = \operatorname{tp}(\langle a, b\rangle, \{e\})$ and

$$r_{ea}^{-1} \circ r_{eb} = s_{ed}$$

for some $d \in A - A[e]$. □

We now fix the notation s for the partial three-place operation in $\mathscr{O}^3$ introduced in Lemma 3.3.

For arbitrary $e \in A - A[\varnothing]$ we denote

$$V_0^e = \{s_{ea} \in T : a \in A - A[e]\}.$$

LEMMA 3.4. *If $d, e \in A - A[\varnothing]$, then*

$$V_0^d = V_0^e.$$

PROOF. By the definition of linearity, for any $a \in A - A[e]$ there exists $b \in A[a, d, e]$ such that $s_{ea} = s_{db}$, and by Lemma 3.1 $a \in A[b, d, e]$. Hence $b \notin A[d]$ and $V_0^e \subseteq V_0^d$. The rest now follows by symmetry. □

Now for arbitrary $e \in A - A[\varnothing]$ we define

$$V_0 = V_0^e, \qquad V = \{t \in T : t \cdot V_0 \cap V_0 \text{ is infinite}\}.$$

LEMMA 3.5. *V is a $\varnothing$-definable strongly minimal group that is a normal subgroup of T, $V \supseteq V_0$. For every $e \in A - A[\varnothing]$ there exists an e-definable bijection $\sigma_e: A - A[e] \to V - V[\varnothing]$.*

REMARK. The group T need not be definable.

PROOF OF THE LEMMA. Since A is strongly minimal,

$$p_{ea} \in V \Leftrightarrow \exists b \in A - A[e, a] \quad p_{ea} \circ s_{eb} = s_{ec} \in V_0$$

for any $p \in \mathscr{O}^3$. Hence $p_{ea} = s_{ec} \circ s_{eb}^{-1}$ and $V_0 \subseteq V$, since if $s_{ea}, s_{eb} \in V_0^e$, $b \in A - A[a, e]$ it follows from Lemma 3.3 that there exists $s_{ec} \in V_0$:

$$s_{ec} = s_{ea}^{-1} \circ s_{eb}, \quad c \in A[e, a, b] - (A[e, a] \cup A[e, b]), \quad s_{ea} \circ s_{ec} = s_{eb} \in V_0;$$

consequently, $s_{ea} \in V$.

V is a group, since for $p_{ea}, r_{eb} \in V$, taking $c \in A - A[e, a, b]$, we obtain

$$r_{eb} \circ s_{ec} = s_{ed} \in V_0, \quad d \in A[e, b, c] - A[e, a], \quad p_{ea} \circ s_{ed} \in V_0,$$

that is, $p_{ea} \circ r_{eb} \in V$. On the other hand, the definition directly yields $p_{ea}^{-1} \in V$.

V is e-definable for any $e \in A - A[\varnothing]$. The elements t of V may be interpreted as triples $\langle e, a, b\rangle \in A^3$ such that $s_{ea} \circ s_{eb}^{-1} = t$, identifying by an e-definable equivalence relation those triples that represent the same translation. When this is done, the condition for an element $t \in V$ to be representable by a triple $\langle e, a, b\rangle$ is expressed, by definition, through the formula

$$R(\{x \in A: (\exists y \in A) s_{ea} \circ s_{eb}^{-1} = s_{ex} \circ s_{ey}^{-1}\}) = 1;$$

V is $\varnothing$-definable, since it is e-definable for all $e \in A - A[\varnothing]$.

To construct the bijection σ_e, we proceed as follows. If $p_{ea} \in V$ is a nontrivial translation, $b \in A - A[e, a]$, then

$$p_{ea} \circ s_{eb} = s_{ec} \in V_0, \quad c \in A - A[e, b], \quad p_{ea} = s_{ec} \circ s_{eb}^{-1} = s_{ed}.$$

This follows from Lemma 3.3. Therefore, setting $\sigma_e(d) = s_{ed}$, we obtain a bijection $A - A[e] \to V - V[\varnothing]$, where $V[\varnothing]$ is the set of trivial translations in V.

σ_e is a partial e-definable map, $\operatorname{Dom}\sigma_e$ is strongly minimal, and consequently $\operatorname{Range}\sigma_e$ is strongly minimal; but $V - \operatorname{Range}\sigma_e$ is finite, since $V - \operatorname{Range}\sigma_e$ is a definable set of rank 0.

V is normal in T. Indeed, if $p_{ec} \in T$, $s_{ea} \circ s_{eb}^{-1} \in V$, and a, b are algebraically independent over e, c, then

$$p_{ec}^{-1} \circ s_{ea} \circ s_{eb}^{-1} \circ p_{ec} = (p_{ec}^{-1} \circ s_{ea} \circ p_{ec}) \circ (p_{ec}^{-1} \circ s_{eb}^{-1} \circ p_{ec}) = r_{ex} \circ r_{ey}^{-1}$$

for some algebraically independent over e elements $x, y \in A$ and $r \in \mathscr{O}^3$. Then by Lemma 3.3, $r_{ex} \circ r_{ey}^{-1} \in V$. This shows that the set

$$t^{-1} \circ V \circ t \cap V$$

is infinite for any $t \in T$. The strong minimality of V now implies that $t^{-1} \circ V \circ t = V$. $\square$

Let us now study the e-definable map σ_e, which is injective on a certain cofinite e-definable subset A_e of A. By adding to A_e a finite set of e-definable imaginary elements, we can construct an e-definable strongly minimal set $\overline{A}_e$ to which σ_e extends as a bijection onto V. We now let $\sigma_e : \overline{A}_e \to V$ denote this bijection.

We define on $\overline{A}_e$ an operation + dependent on σ_e:

$$a + b = c \Leftrightarrow \sigma_e(a) \circ \sigma_e(b) = \sigma_e(c).$$

The expression on the right is a superposition of translations in V. Recall that $(V, \circ)$ is an abelian group, since V is strongly minimal.

If σ_e is independent of e, the group $(\overline{A}_e, +)$ will be $\varnothing$-definable. In general it will be e-definable. Let p_{ea} denote the quasitranslation $x \to a + x$ on $\overline{A}_e$, hence on A. Then for elements a, b independent over e there exists c such that

$$\sigma_e(c) = s_{ec} = p_{ea} \circ p_{eb}^{-1}, \quad \text{i.e., } s_{ec}(x) = (a - b) + x,\ a - b \in \operatorname{dcl}(e, c).$$

We may therefore assume, using the fact that s may be chosen at will, that $a - b = c$, $\sigma_e(c)(x) = c + x$ for $c \in A - A[e]$. The map σ_e thus defined extends uniquely to $\overline{A}_e$:

$$\sigma_e(c)(x) = c + x \quad \text{for all } c \in \overline{A}_e.$$

Note that the translation $\sigma_e(c)$ thus defined acts everywhere on $\overline{A}_e$.

PROPOSITION 3.6. *For every $p \in \mathscr{O}^3$ and $e \in A - A[\varnothing]$, there exists a $\varnothing$-definable automorphism $\overline{p}$ of the abelian group V; the e-definable automorphism $\overline{p}_e$ of the group $\overline{A}_e$ induced by this automorphism is such that for any $a \in A - A[e]$ there exists $c \in \overline{A}_e[e, a]$ satisfying the condition*

$$p_{ea}(x) = \overline{p}_e \cdot x + c,$$

where p_{ea} is a quasitranslation on A, $x \in (A - A[e, a]) \cap \overline{A}_e$, and $\overline{p}_e \cdot x$ denotes the action of $\overline{p}_e$ on x in $\overline{A}_e$.

The set of homomorphisms

$$P = \{\overline{p} : p \in \mathscr{O}^3\} \cup \{\textit{zero homomorphism}\}$$

forms a division ring relative to the operations of addition and superposition of homomorphisms.

If σ_e is independent of e, then the automorphism $\overline{p}_e$ is independent of e and is $\varnothing$-definable.

PROOF. Let $a \in A - A[e]$, $p_{ea} \in T$. Denote by $v^{p_{ea}}$ the conjugation action $p_{ea} \circ v \circ p_{ea}^{-1}$ on an element $v \in V$. Let p_{ea} be nontrivial. Then for arbitrary $b \in A - A[e, a]$, by the definition of s, there exists $d \in A - A[e]$

such that $p_{ea} \circ s_{ed} = p_{eb}$. Hence it follows that $v^{p_{ea}} = v^{p_{eb}}$, and therefore $v^{p_{ea}} \in \mathrm{dcl}(v, e)$. On the other hand, for any $e' \in A - A[\varnothing]$ there exists $a' \in A - A[e']$ such that $p_{e'a'} = p_{ea}$. Therefore $v^{p_{ea}} \in \mathrm{dcl}(v)$. In other words, the automorphism $\overline{p}: v \to v^{p_{ea}}$ depends only on p and is $\varnothing$-definable. This fact is obvious for a trivial translation p_{ea}.

The automorphism $\overline{p}_e$ induced by the automorphism $\overline{p}$ on $\overline{A}_e$ is defined by the following condition:

$$s_{e\overline{p}_e \cdot x} = s_{ex}^{\overline{p}} \quad \text{or} \quad s_{e\overline{p}_e \cdot x} = p_{ea} \circ s_{ex} \circ p_{ea}^{-1},$$

whence

$$s_{e\overline{p}_e \cdot x} \circ p_{ea} = p_{ea} \circ s_{ex}.$$

Using the fact that $s_{ex}(y) = x + y$, we see that for $y \in A - A[e, a, x]$

$$\overline{p}_e \cdot x + p_{ea}(y) = p_{ea}(x + y).$$

Denoting $x + y = z$, we obtain

$$\overline{p}_e \cdot (z - y) + p_{ea}(y) = p_{ea}(z);$$

but $\overline{p}_e$ is an automorphism, and so

$$\overline{p}_e \cdot z - \overline{p}_e \cdot y + p_{ea}(y) = p_{ea}(z),$$

whence

$$\overline{p}_e \cdot z - p_{ea}(z) = \overline{p}_e \cdot y - p_{ea}(y) = c,$$

where $c \in A[e, a]$. Thus,

$$p_{ea}(y) = \overline{p}_e \cdot y + c.$$

Finally, we show that P is closed under addition. To this end we observe that if $\overline{p}, \overline{r} \in P$, then the map $g: \overline{A}_e \to \overline{A}_e$ defined by

$$g(y) = \overline{p}_e \cdot y + \overline{r}_e \cdot y$$

is either the zero homomorphism or, by Lemma 1.1, a quasitranslation of A and an automorphism of $\overline{A}_e$. Consider the second case. As already proved, for all $y \in A - A[e]$

$$g(y) = \overline{r}_e \cdot y + c, \qquad c \in A[e].$$

The fact that g is a homomorphism implies that $c = 0$, whence it follows that $\overline{p}_e + \overline{r}_e = \overline{u}_e$, and so also $\overline{p} + \overline{r} = \overline{u} \in P$.

That P is closed under superposition is a direct corollary of the fact that T is closed under superposition.

COROLLARY 3.7. *The group V, considered relative to the action of the division ring P, is a $\varnothing$-definable vector space. This vector space is isomorphic to the e-definable vector space $\overline{A}_e$ over P. In that case, for any $x, x_1, \dots, x_n \in \overline{A}_e$,*

$$x \in \mathrm{acl}(x_1, \dots, x_n, e) \Leftrightarrow \exists \overline{p}^1, \dots, \overline{p}^n \in P, c \in \overline{A}_e[e]: \quad x = \overline{p}_e^1 \cdot x_1 + \cdots + \overline{p}_e^n \cdot x_n + c.$$

PROOF. Only the last assertion, and even that only in the direction $\Rightarrow$, needs a proof.

We may assume that $x_1, \dots, x_n, e$ are algebraically independent. Since

$$\overline{A}_e[x_1, \dots, x_n, e] = \mathrm{dcl}(x_1, \dots, x_n, e) \cap \overline{A}_e,$$

it follows from $x \in \overline{A}_e[x_1, \dots, x_n, e]$ that

$$x = p_{ey}(x_n), \quad p \in \mathscr{O}^3, \quad y \in \overline{A}_e[x_1, \dots, x_{n-1}, e].$$

Hence,

$$x = \overline{p}_e^n \cdot x_n + z, \qquad z \in \overline{A}_e[x_1, \dots, x_{n-1}, e],$$

and the required assertion now follows by induction. □

Throughout the sequel, we shall use additive notation for the vector space V.

LEMMA 3.8. *For any* $v, v_1, \dots, v_n \in V$,

$$v \in \mathrm{acl}(v_1, \dots, v_n) \Leftrightarrow \exists \overline{p}^1, \dots, \overline{p}^n \in P, c \in V[\varnothing]:$$
$$v = \overline{p}^1 \cdot v_1 + \dots + \overline{p}^n \cdot v_n + c.$$

PROOF. We use the previous assertion and the fact that σ_e is a bijection. We have

$$v \in \mathrm{acl}(v_1, \dots, v_n) \Rightarrow \exists \overline{p}^1, \dots, \overline{p}^n \in P, c \in V[e]:$$
$$v = \overline{p}^1 \cdot v_1 + \dots + \overline{p}^n \cdot v_n + c.$$

Since the left-hand side is independent of e, it follows that c is independent of e, i.e., $c \in V[\varnothing]$. The reverse implication is obvious.

REMARK 3.9. If σ_e is independent of e, then V is isomorphic to $\overline{A} = \overline{A}_e$ by means of a $\varnothing$-definable isomorphism; hence the above lemma describes algebraic closure in $\overline{A}$ as well.

We now study the case in which σ_e depends on e, i.e., $s_{ea} \neq s_{e'a}$ for algebraically independent $e, e', a \in A$. Since V is strongly minimal, this implies that any element v of $V - V[\varnothing]$ may be expressed as $v = s_{ab}$, where a is an arbitrary element of $A - A[\varnothing]$, and b is a suitable element of $A - A[a]$. Since s_{ab} is a nontrivial translation, it follows that $b \in \mathrm{acl}(v, a)$. Therefore, there is no element x in $A - A[\varnothing]$ such that $v \in \mathrm{acl}(x)$, because otherwise, taking $a = x$, we would have $b \in \mathrm{acl}(a)$. Since A is strongly minimal, this implies $\mathrm{tp}(x, v) = \mathrm{tp}(x', v)$ for all $x, x' \in A - A[\varnothing]$. Since the partial action vx is defined for $x \in A - A[v]$, and moreover $vx \in A - A[v]$, the action of any v in V is defined on the whole of $A - A[\varnothing]$. Consider the following x-definable set (the V-orbit of x):

$$Vx = \{vx : v \in V\} \quad \text{for } x \in A - A[\varnothing].$$

By our previous remark, Vx is an infinite subset of $A - A[\varnothing]$. Therefore, for any $x_1, x_2 \in A - A[\varnothing]$, the V-orbits Vx_1 and Vx_2 intersect; hence they coincide, and so Vx is a $\varnothing$-definable set, $Vx = A - A[\varnothing]$, $A[\varnothing]$ is finite, and the map $v \to vx$ is a bijection of the vector space V onto $A - A[\varnothing]$. For fixed $x \in A - A[\varnothing]$, a translation is uniquely determined by the value of $y = vx \in A - A[\varnothing]$; we will therefore use the notation $\overrightarrow{xy} = v$. It is easy to see that

$$\overrightarrow{xy} + \overrightarrow{yz} + \overrightarrow{zx} = \overrightarrow{xx} = 0.$$

This shows that the $\varnothing$-definable map $\langle x, y\rangle \to \overrightarrow{xy}$ gives $A - A[\varnothing]$ the structure of the affine space associated with the vector space V.

Let us assume now that

$$x, x_0, \dots, x_n \in A - A[\varnothing] \quad \text{and} \quad x \in \operatorname{acl}(x_0, \dots, x_n).$$

Then $\overrightarrow{xx_0} \in \operatorname{acl}(\overrightarrow{x_1x_0}, \dots, \overrightarrow{x_nx_0})$ in V, and therefore, by Lemma 3.8, there exist $\overline{p}^1, \dots, \overline{p}^n \in P$ and $c \in V[\varnothing]$ such that

$$\overrightarrow{xx_0} = \overline{p}^1 \cdot \overrightarrow{x_1x_0} + \dots + \overline{p}^n \cdot \overrightarrow{x_nx_0} + c.$$

The converse is obviously also true, so that the last relation implies $x \in \operatorname{acl}(x_0, \dots, x_n)$. □

From the fact just proved, Lemma 3.8, Remark 3.9, Proposition 1.4 and Proposition 3.5.7, there follows a description of the strongly minimal quasialgebras.

THEOREM 3.10. *For any strongly minimal quasialgebra A, one of the following statements is true:*

(1) *There exists a group T acting on $A - A[\varnothing]$ in such a way that the map $a \to ta$, where $t \in T$, is the restriction of a map which is $\varnothing$-definable in $\mathfrak{M}$ and is a bijection of $A - A[\varnothing]$ onto itself without fixed points.*

Every n-place relation $Q(v_1, \dots, v_n)$ which is $\varnothing$-definable in $\mathfrak{M}$, restricted to $A - A[\varnothing]$, is a Boolean combination of relations $v_i = tv_j$, where $i, j \in \{1, \dots, n\}$, $t \in T$.

(2) *There exists a $\varnothing$-definable vector space $\overline{A}$ over a division ring K such that $\overline{A} \dot{-} A$ is finite. Any n-place relation $Q(v_1, \dots, v_n)$ on $\overline{A}$ which is $\varnothing$-definable in $\mathfrak{M}$ is a Boolean combination of relations $k_1v_1 + \dots + k_nv_n + c = 0$, where $k_1, \dots, k_n \in K$, $c \in \overline{A}[\varnothing]$.*

(3) *$A[\varnothing]$ is finite, one can define on $A - A[\varnothing]$ a $\varnothing$-definable structure of an affine space over a division ring K, any n-place relation $Q(v_0, \dots, v_{n-1})$ on $A - A[\varnothing]$, $\varnothing$-definable in $\mathfrak{M}$, is a Boolean combination of relations of the form*

$$k_1\overrightarrow{v_0v_1} + \dots + k_{n-1}\overrightarrow{v_0v_{n-1}} + c = 0,$$

where $k_1, \dots, k_{n-1} \in K$, and c is a $\varnothing$-definable parallel translation of the affine space.

CHAPTER V

Global Structure of Uncountably Categorical Structures

In this chapter we consider the global structure of uncountably categorical structures. By this we mean a description of the structure as a system with distinguished definable subsets, in particular, in terms of algebraic closure. The "building blocks" for such a description will be strongly minimal sets. An essential role in the description of global structure will be played by the groups definable in uncountably categorical structures. In those cases where the structure of these groups is well known, one can obtain a fuller description. In concentrated form, the global structure is described by the following theorems.

THEOREM 0.1 (Ladder Theorem, general case). *Let* M, M_0 *be* $\varnothing$*-definable sets in an uncountably categorical structure,* M_0 *strongly minimal. Then there exists a finite sequence of* $\varnothing$*-definable sets* $M_0, M_1, \dots, M_k$ *ending in* M *such that for each* i $(0 \le i < k)$ *and each* M_i*-atom* $A \subseteq M_{i+1}$, *the group* $\mathrm{Gal}(A, M_i)$ *of all* M_i*-elementary automorphisms of the set* A *is definable together with its action on* A, *and moreover,* $\mathrm{Gal}(A, M_i) \subseteq \mathrm{dcl}(M_i)$.

It is a useful observation that, since $\mathrm{Gal}(A, M_i)$ acts transitively on A, it follows that $R(A) \le R(\mathrm{Gal}(A, M_i))$. In particular, $M_{i+1} \subseteq \mathrm{acl}(M_i)$ if and only if the groups $\mathrm{Gal}(A, M_i)$ are finite for all M_i-atoms.

THEOREM 0.2 (Ladder Theorem, module-like structures). *Let* M, M_0 *be* $\varnothing$*-definable sets in a module-like uncountably categorical structure,* $R(M_0) \ge 1$. *Then there exists a finite sequence of* $\varnothing$ *definable sets* $M_0, \dots, M_k$, *ending in* M, *such that for every* i $(0 \le 1 < k)$ *and every* M_i*-atom* $A \subseteq M_{i+1}$, *either* $R(A) = 0$ *or* $R(A) = R(\mathrm{Gal}(A, M_i)) = 1$ *and* $\mathrm{Gal}(A, M_i)$ *is an abelian group which is* X*-definable together with its action on* A *for any* X *such that* A *is* X*-definable.*

For structures of discrete type the "ladder" degenerates.

THEOREM 0.3 (Ladder Theorem, structures of discrete type). *For any* $\varnothing$*-definable infinite set* M_0 *in a structure of discrete type,* $\mathrm{acl}(M_0)$ *contains all elements of the structure.*

§1. Equivalence generated by the relationship between definable subsets

In this section we consider a fixed pair of $\varnothing$-definable sets A and N such that A is an atom over N, A and N are infinite. The main result is the construction of a certain $\varnothing$-definable equivalence relation $E_{A,N}$ that characterizes the relationship of A to N.

Given a finite subset Z of the structure, define

$$\mathrm{rd}_Z(A, N) = \min\{\mathrm{RD}(x, Y \cup N \cup Z) : x \in A,\ Y \subseteq A,\ x \text{ is independent of } Y \text{ over } Z\}.$$

We define

$$\mathrm{rd}(A, N) = \mathrm{rd}_{\varnothing}(A, N);$$

let $r_Z(A, N)$, $d_Z(A, N)$ denote the first and second coordinates of $\mathrm{rd}_Z(A, N)$ respectively.

LEMMA 1.1. $r(A, N) < R(A)$.

PROOF. By Theorem 1.2.3, there exists a Y-definable stratification ψ of A with respect to B of rank less than $R(A)$, where $Y \subseteq A \cup N$. This means that for any x in A there exists $y \in N$ such that $x \in \psi(y, \mathfrak{M})$ and $R(\psi(y, \mathfrak{M})) < R(A)$. Hence, for any $x \in A$

$$R(A) > R(x, Y' \cup N), \quad \text{where } Y' = Y \cap A,$$

which proves the lemma. □

LEMMA 1.2. *If* $Z \subseteq N$, *then* $\mathrm{rd}_Z(A, N) = \mathrm{rd}(A, N)$.

PROOF. Let $x \in A$ and $Y \subseteq A$. In view of the obvious inequalities

$$R(A) = R(x, Z) = R(x, \varnothing) \geq R(x, Y) \geq R(x, Y \cup Z),$$

it follows from the condition $R(x, Y \cup Z) = R(x, Z)$ that $R(x, Y) = R(x, \varnothing)$. That is, if x, Y are independent over Z, then they are independent over $\varnothing$. Therefore, $\mathrm{rd}_Z(A, N) \geq \mathrm{rd}(A, N)$.

Now let $\mathrm{RD}(x, Y \cup N) = \mathrm{rd}(A, N)$ and let x, Y be independent over $\varnothing$, i.e., $R(x, Y) = R(x, \varnothing) = R(A)$. Let p be an extension of the type $\mathrm{tp}(x, Y)$, complete over $Z \cup Y$, such that $R(p) = R(x, Y)$, and choose an element x' realizing the type p. Then $\mathrm{tp}(x, Y) = \mathrm{tp}(x', Y)$ and therefore $\mathrm{RD}(x', Y \cup N) = \mathrm{rd}(A, N)$. On the other hand x', Y are independent over Z, and therefore

$$\mathrm{RD}(x', Y \cup Z \cup N) \geq \mathrm{rd}_Z(A, N).$$

Since $Z \cup N = N$, it follows that $\mathrm{rd}(A, N) \geq \mathrm{rd}_Z(A, N)$. □

LEMMA 1.3. *There exists a tuple* $\overline{a}$ *of elements of* A *such that for all* x *in* A *that are independent of* $\overline{a}$,

$$\mathrm{RD}(x, |\overline{a}| \cup N) = \mathrm{rd}(A, N).$$

PROOF. Let $\{p_i : i \leq m\}$ be all types complete over A whose rank equals $R(A)$, $A \in p_i$. There exists a finite subset $Y_0 \subseteq A$ such that for all $i \leq m$ $p_i|_{Y_0}$ has a unique complete extension p_i of rank $R(A)$ over A. Let $x_1, \dots, x_m$ realize the types $p_1|_{Y_0}, \dots, p_m|_{Y_0}$, respectively. Since A is an atom, for each x_i there exists a finite set $Y_i \subseteq A$ such that

$$\mathrm{RD}(x_i, Y_i \cup N) = \mathrm{rd}(A, N).$$

Note that by the definition, if $Y \supseteq Y_i$ and x_i is independent of Y, then

$$\mathrm{RD}(x_i, Y \cup N) = \mathrm{rd}(A, N)$$

as well. Set $Y = Y_0 \cup Y_1 \cup \dots \cup Y_m$.

Let x be independent of Y and $x \in A$. Then $R(x, Y) = R(x, \varnothing) = R(A)$ and therefore $\mathrm{tp}(x, Y)$ is an extension of the type $p_i|_{Y_0}$ for some $i \leq m$. Hence

$$\mathrm{tp}(x, Y) = \mathrm{tp}(x_i, Y),$$

and this implies the statement of the lemma if $\overline{a}$ denotes Y, ordered in some way. □

Fix a tuple $\overline{a}$ of elements of A as stipulated in Lemma 1.3.

Let Y be a finite set, x an element. Set

$$x^Y = \{z : \mathrm{tp}(z, Y \cup N) = \mathrm{tp}(x, Y \cup N)\}.$$

If Y consists of a single element y, we shall write x^y instead of $x^{\{y\}}$.

LEMMA 1.4. *For any x and Y,*
(a) *x^Y is definable over $Y \cup N$;*
(b) *x^Y is definable over $Y \cup \{x\}$;*
(c) *if $\mathrm{tp}(\overline{y}, \varnothing) = \mathrm{tp}(\overline{a}, \varnothing)$, $x \in A$ and x is independent of $\overline{y}$, then*

$$\mathrm{RD}(x^{\overline{y}}) = \mathrm{rd}(A, N).$$

PROOF. (a) is a corollary of the fact that an uncountably categorical structure is atomic over every finite definable subset of itself.

(b) follows from (a) and Proposition 1.1.1.

(c) follows from (a) and the definition of $\overline{a}$. □

LEMMA 1.5. *Let $\mathrm{tp}(\overline{y}_1, \varnothing) = \mathrm{tp}(\overline{y}_2, \varnothing) = \mathrm{tp}(\overline{a}, \varnothing)$, $x_1, x_2 \in A$, let x_1 be independent of $\overline{y}_1$, x_2 independent of $\overline{y}_2$, and $x_1^{\overline{y}_1} \cap x_2^{\overline{y}_2} \neq \varnothing$. Then $x_1^{\overline{y}_1} \sqsupset x_2^{\overline{y}_2}$.*

PROOF. Choose an arbitrary element x_3 in $x_1^{\overline{y}_1} \cap x_2^{\overline{y}_2}$. Then by definition x_3 is independent of $\overline{y}_1$, and x_3 is independent of $\overline{y}_2$.

Choose $\overline{y}_3$ to realize the type $\mathrm{tp}(\overline{a}, \varnothing)$ in such a way that

$$R(\overline{y}_3, \{x_3, \overline{y}_1, \overline{y}_2\}) = R(\overline{a}, \varnothing) = R(\overline{y}_3, \varnothing).$$

Calculating ranks, we see from this that $\bar{y}_3$ is independent of x_3 over $\bar{y}_1$, and that $\bar{y}_3$ is independent of x_3 over $\bar{y}_2$. Hence

$$R(x_3, \{\bar{y}_3, \bar{y}_1\}) = R(x_3, \{\bar{y}_1\}), \qquad R(x_3, \{\bar{y}_3, \bar{y}_2\}) = R(x_3, \{\bar{y}_2\}),$$

and since $R(x_3, \{\bar{y}_1\}) = R(x_3, \varnothing) = R(x_3, \{\bar{y}_2\})$, it follows that x_3 is independent of $\bar{y}_1{}^\frown\bar{y}_3$ and x_3 is independent of $\bar{y}_2{}^\frown\bar{y}_3$. By Lemma 1.4 $x_3^{\bar{y}_1} \cap x_3^{\bar{y}_3}$ is definable over $\bar{y}_1{}^\frown\bar{y}_3$, but since $\bar{y}_1{}^\frown\bar{y}_3$ is a tuple of elements from A and x_3 is independent of $\bar{y}_1{}^\frown\bar{y}_3$, it follows that

$$\mathrm{RD}(x_3^{\bar{y}_1} \cap x_3^{\bar{y}_3}) \geq \mathrm{rd}(A, N).$$

This, together with the statement of Lemma 1.4(c), implies that $x_3^{\bar{y}_1} \sqsubset x_3^{\bar{y}_3}$. Similarly one proves that $x_3^{\bar{y}_2} \sqsubset x_3^{\bar{y}_3}$. Hence, $x_3^{\bar{y}_1} \sqsubset x_3^{\bar{y}_2}$. Since $x_3^{\bar{y}_1} = x_1^{\bar{y}_1}$, $x_3^{\bar{y}_2} = x_2^{\bar{y}_2}$, the lemma is proved. □

We now define a two-place relation $E_{A,N}$ on A as follows. $x_1 E_{A,N} x_2 \Leftrightarrow$ there exists $\bar{y}$ realizing $\mathrm{tp}(\bar{a}, \varnothing)$ such that x_1 is independent of $\bar{y}$, x_2 is independent of $\bar{y}$ and $x_1^{\bar{y}} = x_2^{\bar{y}}$.

LEMMA 1.6. *$E_{A,N}$ is a $\varnothing$-definable equivalence relation on A.*

PROOF. For convenience, we write E for $E_{A,N}$; E is obviously reflexive and symmetric. We must prove that E is transitive.

Suppose that $x_1 E x_2$ and $x_2 E x_3$ for some $x_1, x_2, x_3 \in A$. This means that for some $\bar{y}_1, \bar{y}_2$ realizing $\mathrm{tp}(\bar{a}, \varnothing)$, $x_1^{\bar{y}_1} = x_2^{\bar{y}_1}$, $x_2^{\bar{y}_2} = x_3^{\bar{y}_2}$, x_1 and x_2 are independent of $\bar{y}_1$, x_2 and x_3 are independent of $\bar{y}_2$. Let us choose $\bar{y}_3$ so that $\mathrm{tp}(\bar{y}_3, \varnothing) = \mathrm{tp}(\bar{a}, \varnothing)$ and x_1, x_2, and x_3 are independent of $\bar{y}_3$. Then, by Lemma 1.5, $x_1^{\bar{y}_3} \sqsubset x_1^{\bar{y}_1}$ and $x_3^{\bar{y}_3} \sqsubset x_3^{\bar{y}_2}$. Noting that $x_2 \in x_1^{\bar{y}_1} \cap x_3^{\bar{y}_2}$, we infer from Lemma 1.5 that $x_1^{\bar{y}_1} \sqsubset x_3^{\bar{y}_2}$. Hence it follows that $x_1^{\bar{y}_2} \cap x_3^{\bar{y}_3} \neq \varnothing$, and this can be true only if $x_1^{\bar{y}_3} = x_3^{\bar{y}_3}$; in particular, $x_1 E x_3$.

We now prove that E is $\varnothing$-definable. Suppose that x_0 does not depend on $\bar{a}$, $x_0 \in A$. Since the structure is prime and atomic over N, there exists a formula $\psi(\bar{v}, u) \in F(N)$ defining the type $\mathrm{tp}(\bar{a}{}^\frown x_0, N)$. Denote

$$\chi(w_1, w_2) = (\exists u)(\exists \bar{v}_1)(\exists \bar{v}_2)(\psi(\bar{v}_1, w_1) \,\&\, \psi(\bar{v}_2, w_2) \\ \&\, \psi(\bar{v}_1, u) \,\&\, \psi(\bar{v}_2, u)).$$

The formula χ defines the equivalence relation E. Indeed, if $\mathfrak{M} \models \chi(x_1, x_2)$ for some $x_1, x_2 \in A$, then there exist x_3 and $\bar{y}_1, \bar{y}_2$ such that

$$\mathrm{tp}(\bar{y}_1{}^\frown x_1, N) = \mathrm{tp}(\bar{y}_2{}^\frown x_2, N) \\ = \mathrm{tp}(\bar{y}_1{}^\frown x_3, N) = \mathrm{tp}(\bar{y}_2{}^\frown x_3, N) = \mathrm{tp}(\bar{a}{}^\frown x_0, N).$$

In particular, x_1 and x_3 are independent of $\bar{y}_1$, x_2 and x_3 of $\bar{y}_2$, and we have $x_1 E x_3$ and $x_2 E x_3$. Since E is an equivalence relation, we obtain $x_1 E x_2$.

Now suppose, conversely, that $x_1 E x_2$ for some $x_1, x_2 \in A$. By definition, there exists $\overline{y}$ such that x_1 and x_2 are independent of $\overline{y}$, $\mathrm{tp}(\overline{y}, \varnothing) = \mathrm{tp}(\overline{a}, \varnothing)$, and $x_1^{\overline{y}} = x_2^{\overline{y}}$. Since A is an atom over N and $\mathfrak{M} \models \psi(\overline{a}, x_0)$, it follows that $\mathfrak{M} \models \psi(\overline{y}_1, x_1) \& \psi(\overline{y}_2, x_2)$ for some $\overline{y}_1, \overline{y}_2$. By Lemma 1.5,

$$x_1^{\overline{y}_1} \sqsubset x_1^{\overline{y}} = x_2^{\overline{y}} \sqsubset x_2^{\overline{y}_2},$$

in particular, $x_1^{\overline{y}_1} \cap x_2^{\overline{y}_2} \neq \varnothing$, i.e., $\mathfrak{M} \models (\exists u)(\psi(\overline{y}_1, u) \& \psi(\overline{y}_2, u))$, and $\mathfrak{M} \models \chi(x_1, x_2)$.

We have thus shown that E is definable by χ. It is easily seen from the definition that E is stable under all automorphisms, whence, in view of Proposition 1.1.1, we infer that E is $\varnothing$-definable. □

Lemma 1.7. *Let $x \in A$, $\mathrm{tp}(\overline{y}, \varnothing) = \mathrm{tp}(\overline{a}, \varnothing)$, and let x be independent of $\overline{y}$. Then*

$$xE_{A,N} \sqsubset x^{\overline{y}};$$

in particular, for all $x \in A$,

$$\mathrm{RD}(xE_{A,N}) = \mathrm{rd}(A, N).$$

Proof. Denote $E = E_{A,N}$. Since $x^{\overline{y}} \subseteq xE$, it follows by Lemma 1.4(c) that $\mathrm{RD}(xE) \geq \mathrm{rd}(A, N)$.

Consider an arbitrary finite subset $\{x_1, \ldots, x_k\}$ of xE. By definition, there exist $\overline{y}_1, \ldots, \overline{y}_k$ such that for all $i \in \{1, \ldots, k\}$, $x_i^{\overline{y}_i} = x^{\overline{y}_i}$, $\mathrm{tp}(\overline{y}_i, \varnothing) = \mathrm{tp}(\overline{a}, \varnothing)$ and x_i is independent of $\overline{y}_i$. Choose $\overline{y}'$ so that $\mathrm{tp}(\overline{y}', \varnothing) = \mathrm{tp}(\overline{a}, \varnothing)$ and $\{x, x_1, \ldots, x_k\}$ is independent of $\overline{y}'$. By Lemma 1.5, $x_i^{\overline{y}'} \sqsubset x_i^{\overline{y}_i} = x^{\overline{y}_i} \sqsubset x^{\overline{y}'}$. Hence $x_i^{\overline{y}'} \cap x^{\overline{y}'} \neq \varnothing$, that is, $x_i^{\overline{y}'} = x^{\overline{y}'}$, whereby we have proven that $\{x_1, \ldots, x_k\} \subseteq x^{\overline{y}'}$ for some $\overline{y}'$ such that $\mathrm{tp}(\overline{y}', \varnothing) = \mathrm{tp}(\overline{a}, \varnothing)$. By Proposition 1.2.7, $x^{\overline{y}'} \sqsupset xE$ for some $\overline{y}'$ realizing the type $\mathrm{tp}(\overline{a}, \varnothing)$. Hence

$$\mathrm{RD}(xE) = \mathrm{rd}(A, N), \qquad x^{\overline{y}} \sqsubset xE. \ \square$$

§2. Definable groups of automorphisms

In this section we study the case when $\mathrm{rd}(A, N) = \langle 0, 1 \rangle$. We shall show that in this case the group of all bijections of A elementary over N is definable.

All the notation and conventions of the preceding section remain in force. We also introduce the notation A^* for an N-atom containing the tuple $\overline{a}$, i.e., an N-atom defining the type $\mathrm{tp}(\overline{a}, N)$. Generally speaking, A^* is Z-definable for some finite $Z \subseteq N$, but to simplify the notation we shall assume, enriching the structure by the constants Z, that $Z = \varnothing$. It is important to note that, by Lemma 1.2, $\mathrm{rd}(A, N)$ remains unchanged when the signature is so enriched.

LEMMA 2.1. *If A is infinite, the following conditions are equivalent*:
(1) $\mathrm{rd}(A, N) = \langle 0, 1\rangle$;
(2) $A \subseteq \mathrm{dcl}(|\overline{a}| \cup N)$;
(3) $A^* \subseteq \mathrm{dcl}(|\overline{a}| \cup N)$;
(4) *for some finite* Z, $A \subseteq \mathrm{dcl}(Z \cup N)$.

PROOF. We first observe that (2) and (3) are equivalent because $A^* \subseteq A^k$, and the projection of A^* onto the first coordinate is precisely A. It is also obvious that (2)$\Rightarrow$(1) and (2)$\Rightarrow$(4).

(4) $\Rightarrow$ (1). Let $\overline{z}$ be some ordering of Z and B an $A \cup N$-atom that contains $\overline{z}$, with B Y-definable for some finite $Y \subseteq A \cup N$. We claim that for any automorphism α which is the identity on $Y \cup N$, for any $x \in A$, $\alpha(x) = x$, whence, by the Criterion 1.1.1, it will follow that the set $\{x\}$ is $Y \cup N$-definable, and so $\mathrm{rd}(A, N) = \langle 0, 1\rangle$. Thus, let $\alpha|_{Y\cup N} = \mathrm{id}|_{Y\cup N}$. Since B is an $A \cup N$-atom, it follows that if $\alpha(\overline{z}) \neq \overline{z}$, then there exists an automorphism α', identical on $A \cup N$, such that $\alpha \circ \alpha'(\overline{z}) = \overline{z}$, where it is obvious that $\alpha \circ \alpha'$ coincides with α on $A \cup N$. We may therefore assume that $\alpha(\overline{z}) = \overline{z}$. But then $\alpha(x) = x$, since $\{x\}$ is $Z \cup N$-definable.

(1) $\Rightarrow$ (2). Indeed, let (1) be true and suppose that x is independent of $\overline{a}$; then $x \in \mathrm{dcl}(|\overline{a}| \cup N)$. If x' is now an arbitrary element of A, choose an element $\overline{y}$ of A^* that is independent of $\{\overline{a}, x'\}$. Then, first, $x' \in \mathrm{dcl}(|\overline{y}| \cup N)$, and, second, $\overline{y} \in \mathrm{dcl}(|\overline{a}| \cup N)$, since every coordinate y_i of $\overline{y}$ is independent of $\overline{a}$. Hence, $x' \in \mathrm{dcl}(|\overline{a}| \cup N)$, $A \subseteq \mathrm{dcl}(|\overline{a}| \cup N)$, and the lemma is proved. □

REMARK. If A is finite, conditions (2), (3), and (4) of Lemma 2.1 are obviously satisfied, where $\overline{a}$ in (2) and (3) may be the tuple of all elements of A.

Throughout the rest of this section we shall assume that all the conditions of Lemma 2.1 are satisfied, with the exception of (1) if A is finite.

LEMMA 2.2. *There exist k and a formula $\chi(\overline{w}, \overline{v}_1, \overline{v}_2) \in F$ such that*

$$\mathfrak{M} \models (\forall \overline{v}_1 \in A^*)(\forall \overline{v}_2 \in A^*)(\exists \overline{w} \in N^k)(\chi(\overline{w}, \overline{v}_1, \overline{v}_2) \,\&\, (\exists! \overline{u})\chi(\overline{w}, \overline{v}_1, \overline{u})).$$

PROOF. By assumption, for every $\overline{x}$ in A^* there exist a formula $\chi_{\overline{x}}(\overline{w}, \overline{v}_1, \overline{v}_2) \in F$ and a tuple $\overline{y}$ of elements of N such that

$$\mathfrak{M} \models \chi_{\overline{x}}(\overline{y}, \overline{a}, \overline{x}) \,\&\, (\exists! \overline{u})\chi_{\overline{x}}(\overline{y}, \overline{a}, \overline{u}).$$

By the Compactness Theorem, we can choose a finite number of formulas $\chi_i, \dots, \chi_m$ such that for any $\overline{x}$ in A^*, we can take one of the χ_i $(1 \le i \le m)$ as $\chi_{\overline{x}}$. Obviously the formulas χ_i may be chosen pairwise inconsistent, and we can then define χ to be

$$\chi_1(\overline{w}, \overline{v}_1, \overline{v}_2) \vee \cdots \vee \chi_m(\overline{w}, \overline{v}_1, \overline{v}_2).$$

Indeed, by construction,

$$\mathfrak{M} \models (\forall \overline{v}_2)(\exists \overline{w} \in N^k)(\chi(\overline{w}, \overline{a}, \overline{v}_2) \,\&\, (\exists! \overline{u})\chi(\overline{w}, \overline{a}, \overline{u})),$$

and since A^* is an atom, this formula is valid for any $\overline{v}_1 \in A^*$ in place of $\overline{a}$. □

LEMMA 2.3. *There exist a $\varnothing$-definable group G and a $\varnothing$-definable operation*

$$\blacksquare\colon A^* \times G \to A^*,$$

defining the action of G on A^* such that

(a) $G \subseteq \mathrm{dcl}(N)$,

(b) *for any $\overline{x}, \overline{x}' \in A^*$ there exists a unique $g \in G$ such that $\overline{x} \blacksquare g = \overline{x}'$.*

PROOF. Let N^* be a $\varnothing$-definable set and E the equivalence relation on N^* defined by

$$N^* = \{\overline{y} \in N^k : (\exists \overline{v}_1 \in A^*)(\exists \overline{v}_2 \in A^*)(\chi(\overline{y}, \overline{v}_1, \overline{v}_2) \,\&\, (\exists! \overline{u})\chi(\overline{y}, \overline{v}_1, \overline{u}))\},$$
$$\overline{y}_1, E\overline{y}_2 \Leftrightarrow (\forall \overline{v}_1 \in A^*)(\forall \overline{v}_2 \in A^*)(\chi(\overline{y}_1, \overline{v}_1, \overline{v}_2) \leftrightarrow \chi(\overline{y}_2, \overline{v}_1, \overline{v}_2)).$$

Note that any $\overline{y}$ in N^* determines a map

$$g_{\overline{y}}\colon A^* \to A^*,$$

where $g_{\overline{y}}(\overline{v}_1) = \overline{v}_2$ means that $\chi(\overline{y}, \overline{v}_1, \overline{v}_2)$. The domain of definition and range of $g_{\overline{y}}$ both coincide with A^*, since A^* is an N-atom. Obviously, $\overline{y}E\overline{y}_2$ is true when $g_{\overline{y}_1} = g_{\overline{y}_2}$, so we may identify N^*/E with the set of all such maps $g_{\overline{y}}$.

Define

$$G = N^*/E, \qquad \overline{x}_1 \blacksquare g_{\overline{y}} = \overline{x}_2 \Leftrightarrow \chi(\overline{y}, \overline{x}_1, \overline{x}_2).$$

Obviously, $G \subseteq \mathrm{dcl}(N^*) \subseteq \mathrm{dcl}(N)$, which proves (a).

It follows from the definition of χ in Lemma 2.2 that for any $\overline{x}, \overline{x}' \in A^*$ there exists g in G such that $\overline{x} \blacksquare g = \overline{x}'$. If in addition $\overline{x} \blacksquare g' = \overline{x}'$, then, since A^* is an N-atom, $\overline{x} \blacksquare g = \overline{x} \blacksquare g'$ for all $\overline{x}$ in A^*, that is, $g = g'$. This proves condition (b).

We now observe that a superposition of maps g_1, g_2 in G is again a map in G. Indeed, let $(\overline{x} \blacksquare g_1) \blacksquare g_2 = \overline{x}'$. By condition (b), we can find $g \in G$ such that $\overline{x} \blacksquare g = \overline{x}'$, i.e., $(\overline{x} \blacksquare g_1) \blacksquare g_2 = \overline{x} \blacksquare g$. Since A^* is an N-atom, this equality is valid for all $\overline{x} \in A^*$.

Similarly one proves that any map g in G is invertible in G. Therefore G is a group with respect to the operation of superposition. □

REMARK. The group G constructed in Lemma 2.3 acts on A^*, but there is generally no action of G on A.

We now introduce another group, which acts on A and binds A to N.

The *Galois group of A over N* is defined as the group of all bijections of A which are elementary over N. We denote this group by $\mathrm{Gal}(A, N)$.

This definition is legitimate for any sets A, N.

Since A is an N-atom, $\mathrm{Gal}(A, N)$ acts transitively on A. Since $A^* \subseteq A^k$, it follows that $\mathrm{Gal}(A, N)$ also acts on A^*; since A^* is an N-atom, the action of $\mathrm{Gal}(A, N)$ on A^* is transitive, i.e., for any $\overline{x}_1, \overline{x}_2 \in A^*$ there exists $\gamma \in \mathrm{Gal}(A, N)$ such that $\gamma(\overline{x}_1) = \overline{x}_2$.

LEMMA 2.4. *For every* $\overline{x} \in A^*$ *there exists a unique group anti-isomorphism*

$$\pi_{\overline{x}}: G \to \mathrm{Gal}(A, N),$$

defined for all $g \in G$ *by the condition*

$$\overline{x} \blacksquare g = \pi_{\overline{x}}(g)(\overline{x}).$$

If $\overline{x}' = \gamma(\overline{x})$, $\gamma \in \mathrm{Gal}(A, N)$, *then* $\pi_{\overline{x}'}(g) = \gamma \cdot \pi_{\overline{x}}(g) \cdot \gamma^{-1}$.

PROOF. As already pointed out, $\mathrm{Gal}(A, N)$ acts transitively on A^*. Therefore, for every $g \in G$ there exists $\gamma \in \mathrm{Gal}(A, N)$ such that $\overline{x} \blacksquare g = \gamma(\overline{x})$. This γ is uniquely determined, since any element $\overline{x}'$ in A^* may be expressed as $\overline{x} \blacksquare h$ for some $h \in G$ and so

$$\gamma(\overline{x}') = \gamma(\overline{x} \blacksquare h) = \gamma(\overline{x}) \blacksquare h = (\overline{x} \blacksquare g) \blacksquare h. \qquad (*)$$

The fact that γ and the map $\overline{x} \to \overline{x} \blacksquare h$ commute in $(*)$ follows from the fact that γ is elementary over N and the map $\overline{x} \to \overline{x} \blacksquare h$ is defined over N, since $h \in \mathrm{dcl}(N)$.

The invertibility of the map $\pi_{\overline{x}}$ follows from condition (b) of Lemma 2.3.

Now let h be an arbitrary element of G and $\delta = \pi_{\overline{x}}(h)$. We claim that

$$\pi_{\overline{x}}(g \cdot h) = \pi_{\overline{x}}(h) \cdot \pi_{\overline{x}}(g),$$

that is, $\pi_{\overline{x}}$ is an anti-isomorphism. Indeed, we need only show that

$$(\overline{x} \blacksquare g) \blacksquare h = \gamma(\delta(\overline{x})).$$

But this follows at once by evaluating the right-hand side in accordance with $(*)$, taking into account that $\delta(\overline{x}) = \overline{x} \blacksquare h$. □

It follows from Lemma 2.4 that $\mathrm{Gal}(A, N)$ is isomorphic to the group $(G, *)$, where the operation $*$ is antimultiplication on the original group G of Lemma 2.3, i.e., $g * h = h \cdot g$ for all $h, g \in G$. Thus $\mathrm{Gal}(A, N)$ is $\varnothing$-definable as a group without action.

PROPOSITION 2.5. *The group* $\mathrm{Gal}(A, N)$ *is* $\overline{x}$*-definable together with its action on* A *for any* $\overline{x} \in A^*$. *In fact, the set* $\mathrm{Gal}(A, N)$ *is interpreted as* G *and the map*

$$\langle g, y \rangle \to \pi_{\overline{x}}(g)(y) \qquad (g \in G,\ y \in A)$$

from $G \times A$ *to* A, *where* G *and* $\pi_{\overline{x}}$ *are as defined in Lemmas* 2.3 *and* 2.4, *is* $\overline{x}$*-definable. If* $\mathrm{Gal}(A, N)$ *is an abelian group, then its action on* A *is* $\varnothing$*-definable and* $\mathrm{Gal}(A, N)$ *coincides with* G.

PROOF. We shall show that the action of $\mathrm{Gal}(A, N)$ on A^* is $\overline{x}$-definable. Since this action is induced by the action on A, this will easily imply $\overline{x}$-definability of the action on A.

So let $\overline{y}$ be an arbitrary element of A^*. By Lemma 2.3, $\overline{y}$ is uniquely expressible as $\overline{x} \blacksquare h$ for a suitable element h of G. Since the elements of $\operatorname{Gal}(A, N)$ commute with the action of G on A^*, it follows that $\pi_{\overline{x}}(g)(\overline{x} \blacksquare h) = (\pi_{\overline{x}}(g)(\overline{x})) \blacksquare h$, but by definition $\pi_{\overline{x}}(g)(\overline{x}) = \overline{x} \blacksquare g$. Hence

$$\pi_{\overline{x}}(g)(\overline{y}) = \overline{z} \Leftrightarrow (\exists h \in G)(\overline{y} = \overline{x} \blacksquare h \;\&\; \overline{z} = (\overline{x} \blacksquare g) \blacksquare h),$$

that is, the action is $\overline{x}$-definable.

If $\operatorname{Gal}(A, N)$ is abelian, then $\pi_{\overline{x}}(g) = \pi_{\overline{x}'}(g)$ for any $\overline{x}' \in A^*$, which follows from the relation $\pi_{\overline{x}'}(g) = \gamma \cdot \pi_{\overline{x}}(g) \cdot \gamma^{-1}$. Hence the action in this case is $\varnothing$-definable. □

PROPOSITION 2.6. *Let B be an arbitrary $\varnothing$-definable N-atom. If the group $\operatorname{Gal}(B, N)$ is definable together with an action such that $\operatorname{Gal}(B, N) \subseteq \operatorname{dcl}(N)$, then $\operatorname{Gal}(B, N)$ together with its action is $\overline{b}$-definable for some tuple $\overline{b}$ of elements of B and $B \subseteq \operatorname{dcl}(|\overline{b}| \cup N)$.*

PROOF. Let $\operatorname{Gal}(B, N)$ together with its action be X-definable for some finite set X; assume that, under a suitable interpretation, each element of the group is in $\operatorname{dcl}(N)$. Clearly, B is the orbit of an arbitrary element $x_0 \in B$ under the action of $\operatorname{Gal}(B, N)$, and

$$\gamma(x_0) \in \operatorname{dcl}(\{x_0, \gamma\} \cup X) \subseteq \operatorname{dcl}(\{x_0\} \cup X \cup N)$$

for any $\gamma \in \operatorname{Gal}(B, N)$. It may be assumed that $x_0 \in X$, and then $B \subseteq \operatorname{dcl}(X \cup N)$. But this is simply condition (4) of Lemma 2.1 which, according to Proposition 2.5, is a sufficient condition for $\operatorname{Gal}(B, N)$ to be $\overline{b}$-definable together with its action, for $\overline{b} \in B^*$. □

With any subgroup H of $\operatorname{Gal}(A, N)$ we associate two equivalence relations E_H and E_H^*, defined on A and A^*, respectively, by

$$xE_Hy \Leftrightarrow (\exists \gamma \in H)(\gamma(x) = y),$$
$$\overline{x}E_H^*\overline{y} \Leftrightarrow (\exists \gamma \in H)(\gamma(\overline{x}) = \overline{y}).$$

Conversely, given an arbitrary definable equivalence relation E on A or A^*, we can define a subgroup $H_E < \operatorname{Gal}(A, N)$:

$$H_E = \{\gamma \in \operatorname{Gal}(A, N) \colon (\forall v_1, v_2)(v_1 E v_2 \leftrightarrow \gamma(v_1) E \gamma(v_2))\}.$$

PROPOSITION 2.7. (a) *If H is an arbitrary normal subgroup of* $\operatorname{Gal}(A, N)$, *then E_H and E_H^* are N-definable. If at the same time H is $\varnothing$-definable, then E_H and E_H^* are $\varnothing$-definable.*

(b) *If E is an N-definable equivalence relation, then H_E is normal in* $\operatorname{Gal}(A, N)$.

PROOF. (a) It follows from Proposition 2.5 that E_H and E_H^* are definable. Let $y \in A$ and $E = E_H$ or $y \in A^*$, $E = E_H^*$. Then

$$yE = \pi_{\overline{x}}(H)(y).$$

We claim that E is invariant under automorphisms which are elementary over N, i.e., under elements γ of $\mathrm{Gal}(A, N)$. Indeed,

$$\begin{aligned}\gamma(yE) = \gamma(\pi_{\overline{x}}(H)(y)) &= \gamma \cdot \pi_{\overline{x}}(H) \cdot \gamma^{-1}(\gamma(y)) \\ &= \pi_{\overline{x}}(H)(\gamma(y)) = \gamma(y)E,\end{aligned}$$

that is, an equivalence class is mapped to an equivalence class. By Criterion 1.1.1, E is N-definable.

If H is $\varnothing$-definable, then E is $\overline{x}$-definable for any $\overline{x}$ in A^*. Again using the fact that H is a normal subgroup and the formula from Lemma 2.4 relating $\pi_{\overline{x}}$ to $\pi_{\overline{x}'}$, we see that E is invariant under all automorphisms, i.e., E is $\varnothing$-definable.

(b) follows from the invariance of E under $\mathrm{Gal}(A, N)$. □

PROPOSITION 2.8. *If the group* $\mathrm{Gal}(A, N)$ *is abelian, then*

$$A \subseteq \mathrm{dcl}(\{a\} \cup N)$$

for any a *in* A, *so that we may assume* $A^* = A$.

PROOF. By Proposition 2.5, the action of $\mathrm{Gal}(A, N)$ on A is $\varnothing$-definable in our case. Then for any y in A the formula $y = \pi_{\overline{x}}(g)(a)$, which is valid for a suitable $g \in G$, is independent of $\overline{x}$, and therefore

$$y \in \mathrm{dcl}(a, g) \subseteq \mathrm{dcl}(\{a\} \cup N). \quad \square$$

PROPOSITION 2.9. *If* H *is a normal definable subgroup of* $\mathrm{Gal}(A, N)$, *then*

(a) A^*/E_H^* *is a* $\varnothing$*-definable* N*-atom,*

(b) *for any* $a \in A$, *the class* aE_H *is an* $N \cup (A^*/E_H^*)$*-atom,*

(c) $\mathrm{Gal}(A, N \cup (A^*/E_H^*)) = H$,

(d) $\mathrm{Gal}((A^*/E_H^*), N) = \mathrm{Gal}(A, N)/H$,

where the group on the right is the quotient group acting canonically on H*-orbits, i.e., on* A^*/E_H^*.

PROOF. Let us denote $E = E_H$, $E^* = E_H^*$ for brevity. It follows from the definition of E^* that a monomorphism γ in $\mathrm{Gal}(A, N)$ will leave all E^*-equivalence classes fixed if and only if $\gamma \in H$. On the other hand, a monomorphism γ will be an element of $\mathrm{Gal}(A, N \cup (A^*/E^*))$ if and only if γ can be extended to $A \cup N \cup (A^*/E^*)$ by defining it as the identity on $N \cup (A^*/E^*)$. In that case $\gamma \in \mathrm{Gal}(A, N)$ and γ is the identity on E^*-classes. This proves (c).

It follows from (c) that the group of automorphisms coinciding with the identity on $N \cup (A^*/E^*)$ is transitive on aE, since H is transitive. This gives (b).

By Proposition 2.7, E^* is invariant under the action of $\mathrm{Gal}(A, N)$. Therefore, a monomorphism γ in $\mathrm{Gal}(A, N)$ induces on A^* a monomorphism $\hat{\gamma}$ on A^*/E^*. It follows directly from the definition of E^* that the subgroup of $\mathrm{Gal}(A, N)$ consisting of all elements which are the identity on A^*/E^*

is precisely H. Therefore, the kernel of the homomorphism $\gamma \to \hat{\gamma}$ is H. On the other hand, the homomorphism $\gamma \to \hat{\gamma}$ is an epimorphism, since any monomorphism on A^*/E^* which is elementary over N extends to an automorphism of the structure which is elementary over N. This proves (d). Finally, (a) follows from the transitivity of the action of $\mathrm{Gal}(A, N)/H$ on A^*/E^*. □

PROPOSITION 2.10. *Let A, N, N' be $\varnothing$-definable sets, $A' \subseteq A$, $N \subseteq N'$, A an N-atom, A' an N'-atom. Then there exist subgroups H_1, H_2 of* $\mathrm{Gal}(A, N)$, $H_1 \lhd H_2 \lhd \mathrm{Gal}(A, N)$, *such that*

$$\mathrm{Gal}(A', N') \cong H_1/H_2.$$

If $\mathrm{Gal}(A, N)$ *is definable together with its action on A, then the groups H_1, H_2,* $\mathrm{Gal}(A', N')$ *are also definable together with their actions on A, A, A', respectively.*

PROOF. Define

$$H_1 = \mathrm{Gal}(A, N'), \qquad H_2 = \{\gamma \in H_1 : \gamma \text{ is the identity on } A'\}.$$

Let $\eta \in H_1$, $\gamma \in \mathrm{Gal}(A, N)$. We shall assume that η, γ act on N', with η the identity on N'. Since N' is $\varnothing$-definable, $\gamma(N') = N'$, therefore $\gamma^{-1} \cdot \eta \cdot \gamma \in H_1$, i.e., H_1 is normal in $\mathrm{Gal}(A, N)$. It is easily seen from the definition that the map $\eta \to \hat{\eta}$, where $\hat{\eta}$ is the restriction of η to A', is a homomorphism of H_1 onto $\mathrm{Gal}(A', N')$ compatible with the action and the kernel of H_2.

If $\mathrm{Gal}(A, N)$ is definable together with its action, then for any formula $\chi(v_1, \dots v_k) \in F$ there is a definable subgroup

$$H^\chi = \{\gamma \in \mathrm{Gal}(A, N) : (\forall v_1, \dots, v_k \in A) \\ (\chi(v_1, \dots, v_k) \leftrightarrow \chi(\gamma(v_1), \dots, \gamma(v_k)))\}.$$

Obviously, H_1 is the intersection of H^χ over all such formulas, and the intersection is definable. That H_2 is definable as a subgroup of H_1 is obvious. □

§3. Ladder theorems

We now turn to the proof of the Ladder Theorem (general case).

LEMMA 3.1. *Let $R(a, \varnothing) > 1$ and suppose that for any $\varnothing$-definable set N, if $R(N) < R(A, \varnothing)$, then $a \notin$* $\mathrm{acl}(N)$. *Then there exists a $\varnothing$-definable set N such that*

(1) $0 < R(N) < R(a, \varnothing)$;

(2) *there exist $b \in$* $\mathrm{dcl}(a) \cap N$ *and a b-definable set A such that $a \in A$ and A is an N-atom*;

(3) *for any b-definable equivalence relation E_b on A, either the class aE_b is finite, or $R(aE_b) = R(A)$.*

PROOF. Let N_0 be a $\varnothing$-definable infinite set such that $0 < R(N_0) < R(a, \varnothing)$ and $R(a, N_0)$ takes the least possible value. Since an uncountably categorical structure is atomic over any of its definable infinite subsets, there exists an N_0-atom A containing a.

Let $\varphi(\overline{v}, w)$ be a formula without constants such that for some tuple $\overline{c}_0$ in N_0^k, $\varphi(\overline{c}_0, \mathfrak{M}) = A$, and let

$$N_1 = \{\overline{c} \in N_0^k : (\forall \overline{c}' \in N_0^k)\varphi(\overline{c}, \mathfrak{M}) \cap \varphi(\overline{c}', \mathfrak{M}) \neq \varnothing \rightarrow \varphi(\overline{c}, \mathfrak{M}) \subseteq \varphi(\overline{c}', \mathfrak{M})\}.$$

It is clear that $\overline{c}_0 \in N_1$, N_1 is $\varnothing$-definable, and the relation I on N_1 defined by the condition

$$\overline{c} I \overline{c}' \Leftrightarrow \varphi(\overline{c}, \mathfrak{M}) = \varphi(\overline{c}', \mathfrak{M})$$

is an equivalence relation on N_1, and moreover for $\overline{c} \in N_1$

$$a \in \varphi(\overline{c}, \mathfrak{M}) \Leftrightarrow \| \overline{c} \in \overline{c}_0 I.$$

Define $N = N_1/I$, $b = \overline{c}_0 I$. It is easy to see that condition (2) applies and $R(b, \varnothing) < R(a, \varnothing)$, since $b \in \operatorname{dcl}(a)$, $b \in \operatorname{dcl}(N_0)$, but $a \notin \operatorname{acl}(N_0)$. Replacing N if necessary by a $\varnothing$-definable subset of N, we can obtain $R(N) = R(b, \varnothing)$ and consequently condition (1) applies. We note that $R(a, N) = R(a, N_0)$.

Let U be a $\varnothing$-definable set, $A \subseteq U$, and $J \subseteq N \times U$ a $\varnothing$-definable relation such that

$$xJu \Leftrightarrow (\exists \overline{y} \in N_1)(\varphi(\overline{y}, u) \,\&\, x = \overline{y} I).$$

It follows from the construction of N_1 and I that Ju is either a singleton or the empty set.

Let E_x be a formula with parameter x and two free variables which, for $x = b$, defines the equivalence relation E_b on A. Replacing N if necessary by a suitable subset, we may assume that for all $x \in N$ E_x is an equivalence relation on A. We now define an equivalence relation E on J by

$$\langle x, u \rangle E \langle x', u' \rangle \Leftrightarrow x = x' \,\&\, u E_x u',$$

and set

$$N^* = J/E, \qquad b^* = \langle b, a \rangle E.$$

It is easy to see that

$$R(aE) = R(aE_b).$$

We may assume without loss of generality that $R(b^*) = R(N^*)$. It is easy to show that $R(a, \{b^*\}) = R(aE)$, and hence $R(a, N^*) \leq R(aE)$. It follows from the minimality of $R(a, N_0)$ that either $R(b^*, \varnothing) = 0$, or $R(b^*, \varnothing) \geq R(a, \varnothing)$, or $R(aE) = 0$. Since $b \in \operatorname{acl}(a)$, it follows that

$$R(\langle b, a \rangle, \varnothing) = R(a, \varnothing), \qquad R(b^*, \varnothing) = R(A) - R(aE).$$

Hence,

$$R(b^*, \varnothing) = 0 \Leftrightarrow R(aE) = R(A),$$
$$R(b^*, \varnothing) \geq R(a, \varnothing) \Leftrightarrow R(aE) = 0,$$

which proves that $R(aE_b) = 0$, or $R(aE_b) = R(A)$. □

LEMMA 3.2. *Under conditions* (2), (3) *of Lemma* 3.1, *we can find* $\hat{a}$ *and an* N*-atom* $\widehat{A}$ *such that* $\hat{a} \in \widehat{A}$, $\mathrm{acl}(\hat{a}) = \mathrm{acl}(a)$ *and the following condition holds*:

(4) $\mathrm{rd}(\widehat{A}, N) = \langle 0, 1 \rangle$.

PROOF. Let N and b satisfy conditions (2), (3), and let $E_{A,N}$ be the b-definable equivalence relation on A introduced in §1, which we henceforth denote by E_b. By condition (3), all the equivalence classes of E_b are finite.

Let U, J, E, N^* be the same as in the proof of Lemma 3.1, and for $u \in U$ denote

$$\hat{u} = \langle b, u \rangle E, \qquad \widehat{A} = \{\hat{u} : u \in A\};$$

$\widehat{A}$ may be identified naturally with the quotient set A/E_b. Clearly, $\widehat{A}$ is a b-definable subset of N^*, $b \in \mathrm{dcl}(\hat{a})$, and in addition $\widehat{A}$ is an N-atom, $\mathrm{acl}(\hat{u}) = \mathrm{acl}(u)$, since E_b is finite and $b \in \mathrm{dcl}(u)$ for $u \in U$.

By Lemma 1.3, there exists a tuple $\overline{a}$ of elements of A such that if x is independent of $\overline{a}$ over b, then

$$\mathrm{RD}(x, |\overline{a}| \cup N) = \mathrm{rd}(A, N) = \mathrm{RD}(xE_b).$$

Let $\varphi_1(\overline{v})$ be a formula in $F(N \cup \widehat{A})$ which defines a complete type $\overline{a}$ over $N \cup \widehat{A}$ and let $\overline{c}$ be the tuple of the constants of $\widehat{A}$ figuring in φ_1. We may assume that $\hat{a}_i$ appears in $\overline{c}$ if a_i is from $\overline{a}$.

Let $x \in A$, $\hat{x} \in \widehat{A}$, and let $\hat{x}$ be independent of $\overline{c}$ over b. Then x is independent of $\overline{a}$ over b, since $\mathrm{acl}(\overline{x}) = \mathrm{acl}(x)$.

Let $\varphi_2(v_0, \overline{v})$ be a formula in $F(N)$ which defines a complete type $\langle x, \overline{a} \rangle$ over N. Since $R(xE_b) = 0$, it follows that $\varphi_2(v_0, \overline{a})$ is equivalent to the formula $\hat{v}_0 = \hat{x}$ by Lemma 1.7. Therefore,

$$\mathfrak{M} \vDash \varphi_1(\overline{a}) \,\&\, (\forall v_0)(\varphi_2(v_0, \overline{a}) \leftrightarrow \hat{v}_0 = \hat{x}).$$

Since $\varphi_1(\overline{v})$ is an atom over $N \cup \widehat{A}$ and $\hat{x} \in \widehat{A}$, it follows that

$$\mathfrak{M} \vDash (\forall \overline{v})(\varphi_1(\overline{v}) \to (\forall v_0)(\varphi_2(v_0, \overline{v}) \leftrightarrow \hat{v}_0 = \hat{x})).$$

Therefore, the formula

$$(\exists \overline{v})(\varphi_1(\overline{v}) \,\&\, \varphi_2(v_0, \overline{v}))$$

is definable over $|\overline{c}| \cup N$ and equivalent to $\hat{v}_0 = \hat{x}$, i.e., $\hat{x} \in \mathrm{dcl}(|\overline{c}| \cup N)$, $\mathrm{rd}(\widehat{A}, N) = \langle 0, 1 \rangle$. □

LEMMA 3.3. *For any $\varnothing$-definable set M of rank greater than 1, one can find $\varnothing$-definable sets $\widehat{M}$ and N such that*

(a) $M \subseteq \operatorname{acl}(\widehat{M})$;

(b) $R(M) > R(N)$;

(c) *for any N-atom $\widehat{A} \subseteq \widehat{M}$, the group* $\operatorname{Gal}(\widehat{A}, N)$ *is definable together with its action,* $\operatorname{Gal}(\widehat{A}, N) \subseteq \operatorname{dcl}(N)$.

PROOF. We first observe that if $M = M_1 \cup \cdots \cup M_k$ and for every M_i $(1 \le i \le k)$ there exist $\widehat{M}_i$ and N_i satisfying conditions (a), (c) and such that $R(N_i) < R(M)$, then $\widehat{M} = \widehat{M}_1 \cup \cdots \cup \widehat{M}_k$, $N = N_1 \cup \cdots \cup N_k$ satisfy the requirements. Indeed, (a) and (b) are obviously true, while (c) follows from Proposition 2.10.

Thus, by the Compactness Theorem, it will suffice to find, for every $a \in M$, $\varnothing$-definable sets M_a, $\widehat{M}_a$, and N_a for which conditions (a), (b), and (c) are satisfied and $a \in M_a$.

For arbitrary $a \in M$ such that $R(a, \varnothing) = R(M)$, it follows from Lemma 3.2 that we can find a $\varnothing$-definable set N_a, an N_a-atom $\widehat{A}$, and an element $\hat{a} \in \widehat{A}$ such that

$$\operatorname{acl}(\hat{a}) = \operatorname{acl}(a), \quad R(\hat{a}, \varnothing) > R(N_a), \quad \operatorname{rd}(\widehat{A}, N_a) = \langle 0, 1\rangle.$$

The first of these conditions implies $R(a, \varnothing) = R(\hat{a}, \varnothing)$; therefore, $R(a, \varnothing) > R(N_a)$.

Let $\widehat{M}_a$ be some $\varnothing$-definable set containing $\widehat{A}$ and satisfying the condition $R(\widehat{M}_a) = R(\hat{a}, \varnothing)$. Let $\varphi(v, w)$ be a formula in F such that $\varphi(b, \mathfrak{M}) = \widehat{A}$ for some $b \in N_a$. Since $\operatorname{rd}(\widehat{A}, N) = \langle 0, 1\rangle$, it follows that

$$\begin{aligned}\mathfrak{M} \models (\exists x_1, \ldots, x_k \in \varphi(b, \mathfrak{M}))(\forall y \in \varphi(b, \mathfrak{M}))(\exists z_1, \ldots, z_k \in N) \qquad (*)\\ (\chi(x_1, \ldots, x_k, z_1, \ldots, z_k, y) \,\&\, (\exists! v)\chi(x_1, \ldots, x_k, z_1, \ldots, z_k, v)).\end{aligned}$$

In other words, condition $(*)$ guarantees that $\operatorname{dcl}(\{x_1, \ldots, x_k\} \cup N) \supseteq \varphi(b, \mathfrak{M})$, but this implies condition (4) of Lemma 2.1 for any N-atom $B \subseteq \varphi(b, \mathfrak{M})$. We now augment the definition of $\widehat{M}_a$ with the following condition.

$$\widehat{M}_a \subseteq \bigcup\{\varphi(b, \mathfrak{M}) \colon b \in N_a \text{ and } b \text{ satisfies } (*)\}.$$

Condition (c) for N_a and $\widehat{M}_a$ now follows from Proposition 2.6. Set

$$M_a = \bigcup\{\operatorname{acl}(x) \colon x \in \widehat{M}_a\} \cap M.$$

For elements a of M that $R(a, \varnothing) < R(M)$, define M_a to be an arbitrary $\varnothing$-definable set containing a, and define $\widehat{M}_a = N_a = M_a$; then obviously $\operatorname{Gal}(A, N_a) = \mathbb{1}$ for $\widehat{A} \subseteq \widehat{M}_a$.

Thus the required sets M_a, $\widehat{M}_a$, N_a have been found, and the proof of the lemma is complete. □

We now proceed to the proof of Theorem 0.1.

Induction on $R(M)$. We first observe that if $M \subseteq \operatorname{acl}(M')$, then any M'-atom $A \subseteq M$ is finite, and this, as shown in §2, is a sufficient condition for $\operatorname{Gal}(A, M')$ to be definable together with its action, $\operatorname{Gal}(A, M') \subseteq \operatorname{dcl}(M')$. In particular, this implies the truth of the theorem for the case $\operatorname{acl}(M_0) \supseteq M$.

Let $R(M) \leq 1$ and $M \not\subseteq \operatorname{acl}(M_0)$. Then there exists an M_0-atom $A \subseteq M$, $R(A) = 1$. There are only finitely many such atoms, so their union is $\varnothing$-definable and may be assumed equal to M, and on M we have a $\varnothing$-definable equivalence relation E with finitely many classes. Thus M/E is a finite $\varnothing$-definable set, i.e., $M/E \subseteq \operatorname{acl}(\varnothing)$. Define $(M/E) \cup M_0 = M_1$; then every M_1-atom $A \subseteq M$ is b-definable for $b \in M_1$. Using Lemma 3.2, we construct

$$\widehat{M} = \bigcup\{\widehat{A}\colon A \text{ is an } M\text{-atom}, A \subseteq M\}.$$

By construction, $\widehat{M}$ is $\varnothing$-definable, $M \subseteq \operatorname{acl}(\widehat{M})$, $R(M) = R(\widehat{M}) = 1$, and every M_1-atom underlying $\widehat{M}$ has the form $\widehat{A}$, $A \subseteq M$, where A is an M_1-atom, the group $\operatorname{Gal}(\widehat{A}, M_1)$ is definable together with its action, $\operatorname{Gal}(\widehat{A}, M_1) \subseteq \operatorname{dcl}(M_1)$. That is to say: defining $M_2 = \widehat{M}$, $M_3 = M$, we obtain the required sequence M_0, M_1, M_2, M_3.

Finally, if the assertion of the theorem is true for $\varnothing$-definable sets of rank less than $R(M)$ with $R(M) > 1$, then, utilizing the notation of Lemma 3.3, we define

$$M_{n+2} = M, \quad M_{n+1} = \widehat{M}, \quad M_n = N,$$

$M_0, M_1, \ldots, M_{n-1}, M_n$ is a sequence whose existence follows from the inductive hypothesis for N. Then $M_0, M_1, \ldots, M_{n+2}$ is the required sequence for M. □

Theorem 0.3 is a direct corollary of Theorem 0.1 just proved. Indeed, $M_{i+1} \subseteq \operatorname{acl}(M_i)$ for all "rungs" of the ladder constructed, since the groups $\operatorname{Gal}(A, M_i)$, and together with them the atoms $A \subseteq M_i$, are finite by Theorem 2.4.5.

LEMMA 3.4. *Let M, N be $\varnothing$-definable sets in a module-like structure, such that for every N-atom $A \subseteq M$ the group* $\operatorname{Gal}(A, N)$ *is defined together with its action. Then there exists a $\varnothing$-definable set N' such that* $\operatorname{acl}(N') = \operatorname{acl}(N)$ *and for any N'-atom $A \subseteq M$ the group* $\operatorname{Gal}(A, N')$ *is definable together with its action and abelian.*

PROOF. It will suffice to find, for any element a of M, $\varnothing$-definable sets M_a and N'_a such that $a \in M_a \subseteq M$, $\operatorname{acl}(N'_a) = \operatorname{acl}(N)$ and for any N'_a-atom $A \subseteq M_a$ the group $\operatorname{Gal}(A, N'_a)$ is definable together with its action and abelian. By the Compactness Theorem, we can then find finitely many sets $M_{a_1}, \ldots, M_{a_k}$ whose union is M. Then $N' = N'_{a_1} \cup \cdots \cup N'_{a_k}$ will be the required set, by Proposition 2.10.

So let a be an arbitrary element of M and consider the N-atom $A \subset M$ containing a. By assumption, the group $\operatorname{Gal}(A, N)$ is definable together with its action. Let H be the connected component of $\operatorname{Gal}(A, N)$, i.e, the minimal definable normal subgroup such that $\mathrm{Gal}(A, N)/H$ is finite. By Theorem 2.4.1 (groups in module-like structures), H is an abelian group.

Now consider the quotient set A^*/E_H^*. This set is N-definable, the group $\mathrm{Gal}(A, N \cup (A^*/E_H^*))$ coincides with H and A^*/E_H^* is finite, by Proposition 2.9.

Again by Proposition 2.9, the equivalence class aE_H is an $N \cup (A^*/E_H^*)$-atom with abelian group $\mathrm{Gal}(aE_H, \cup(NA^*/E_H^*))$ (Proposition 2.10), which in this case is $\overline{x}$-definable together with its action if aE_H is $\overline{x}$-definable (Proposition 2.5). We also note that, denoting $B = aE_H$, we have $B^* = B$ by Proposition 2.8. Let N_a' be a $\varnothing$-definable overset of $(A^*/E_H^*) \cup N$. Since A^*/E_H^* is a finite N-definable set, we can choose N_a' so that $N_a' \subseteq \mathrm{acl}(N)$, and therefore $\mathrm{acl}(N_a') = \mathrm{acl}(N)$.

Finally, let $\varphi(\overline{v}, w)$ be a formula in F such that $\varphi(\overline{x}, \mathfrak{M}) = aE_H$ for some tuple $\overline{x}$ of elements of N_a'. As remarked previously, if $\overline{v} = \overline{x}$ the following condition applies: $(*)$ There exists an abelian group $G_{\overline{v}}$, acting on $\varphi(\overline{v}, \mathfrak{M})$, such that $G_{\overline{v}} \subseteq \mathrm{dcl}(N_a')$ and for all w_1, w_2 such that $\varphi(\overline{v}, w_1) \,\&\, \varphi(\overline{v}, w_2)$ there exists a unique $g \in G_{\overline{v}}$ for which $w_1 \blacksquare g = w_2$. By the Compactness Theorem, there exists a formula $\psi(\overline{v})$ in F such that for all $\overline{y}$ satisfying ψ, condition $(*)$ is valid. Now set

$$M_a = \bigcup \{\varphi(\overline{y}, \mathfrak{M}) : \mathfrak{M} \models \psi(\overline{y})\}.$$

If A' is an arbitrary N_a'-atom, $A' \subseteq M_a$, then $A' \subseteq \varphi(\overline{y}, \mathfrak{M})$ for $\overline{y}$ satisfying ψ. Consider the abelian group $G_{\overline{y}}$, which acts on $\varphi(\overline{y}, \mathfrak{M})$. Define

$$G = \{g \in G_{\overline{y}} : A' \cap A' \blacksquare g \neq \varnothing\}.$$

Since A' is an N_a'-atom, it follows that $A' \blacksquare g = A'$ for all g in G. By condition $(*)$, G satisfies all the assumptions of Lemma 2.3, provided that the signature is enriched with the constants for $\overline{y}$. Hence, by Proposition 2.5, $\mathrm{Gal}(A', N_a')$ is a $\overline{y}$-definable abelian group. □

LEMMA 3.5. *Let M, N be $\varnothing$-definable sets in a module-like structure, $M \not\subseteq \mathrm{acl}(N)$, and suppose that for every N-atom $A \subseteq M$ the group* $\mathrm{Gal}(A, N)$ *is definable together with its action and abelian. Then there exists a $\varnothing$-definable set $N^0 \supseteq N$ such that*

(a) $\max\{R(x, N^0) : x \in M\} = 1$;

(b) $\max\{R(x, N) : x \in N^0\} < \max\{R(x, N) : x \in M\}$;

(c) *for any N-atom $B \subseteq N^0$, the group* $\mathrm{Gal}(B, N)$ *is definable together with its action and abelian*;

(d) *for any N^0-atom $A \subseteq M$, the group* $\mathrm{Gal}(A, N^0)$ *is definable together with its action and abelian.*

PROOF. As before, it will suffice to show that, for every a in $M - \mathrm{acl}(N)$, we can find $\varnothing$-definable sets $M_a \subseteq M$ and N_a satisfying conditions (a), (b), (c), and (d).

Thus, let $a \in M - \operatorname{acl}(N)$, $a \in A \subseteq M$, A an N-atom. By Theorem 2.4.1, the group $\operatorname{Gal}(A, N)$ contains a definable subgroup H of rank 1. We use Proposition 2.9, bearing in mind that in our case $A^* = A$, and define N_a' to be an arbitrary $\varnothing$-definable set containing $N \cup (A/E_H)$. Then for $x \in A$

$$R(x, N_a') \leq R(xE_H) = R(H) = 1.$$

At the same time, A/E_H is an N-atom with abelian Galois group over N, and $R(A/E_H) < R(A)$, since $R(xE_H) = R(H) = 1$ for all $x \in A$. Let $\varphi(\overline{v}, w)$ be a formula in F such that $\varphi(\overline{x}, \mathfrak{M}) = A/E_H$ for some tuple $\overline{x}$ of elements of N. We use condition $(*)$ and the formula ψ from the proof of Lemma 3.4. Define

$$N_a^0 = \bigcup\{\varphi(\overline{y}, \mathfrak{M}) : \mathfrak{M} \models \psi(\overline{y}),\ R(\varphi(\overline{y}, \mathfrak{M})) < r\},$$

where $r = R(A) = R(a, N)$;

M_a is an arbitrary $\varnothing$-definable set containing a such that

$$\max\{R(x, N_a^0) : x \in M_a\} \leq 1.$$

It is easy to verify that these sets N_a^0, M_a satisfy conditions (a), (b), and (c). Condition (d) follows from 2.10. □

PROOF OF THE LADDER THEOREM FOR MODULE-LIKE STRUCTURES. We use the ladder $M_0, M_1, \ldots, M_k = M$ constructed in Theorem 0.1. By Lemma 3.4, this ladder may be "filled in" in such a way that the groups $\operatorname{Gal}(A, M_i)$ for all $i < k$ ($A \subseteq M_{i+1}$, A an M_i-atom) are all finite or all abelian. Finally, Lemma 3.5 makes it possible, proceeding by induction on

$$\max\{R(x, M_i) : x \in M_i\},$$

to construct the rungs of the required ladder. □

§4. Envelopes and the problem of finite axiomatizability of totally categorical theories

Let X and N be subsets of $\mathfrak{M}$. We shall say that X is *m-elementary relative to N* if, for any set C which is definable relative to an m-element subset $Y \subseteq X$, $C \cap N \neq \varnothing$ implies $C \cap X \neq \varnothing$.

If N is the underlying set of $\mathfrak{M}$, X will be called simply an *m-elementary subset.*

PROPOSITION 4.1. *Let $\mathfrak{X}$ be a substructure of $\mathfrak{M}$ whose underlying set X is $m+k$-elementary. Let $\varphi(\overline{v})$ be any formula with m free variables and k quantifiers. Then for any tuple $\overline{x}$ of length m in X*

$$\mathfrak{X} \models \varphi(\overline{x}) \Leftrightarrow \mathfrak{M} \models \varphi(\overline{x}).$$

PROOF. By induction on k. The assertion is obvious for $k = 0$. If φ has $k \geq 1$ quantifiers, we may assume that the left-most one is $\exists$, i.e., $\varphi = (\exists w)\psi$, where ψ is a formula with $m+1$ variables and $k-1$ quantifiers.

By the inductive hypothesis, if $\mathfrak{X} \models \varphi(\overline{x})$, then $\mathfrak{M} \models \varphi(\overline{x})$. But if $\mathfrak{M} \models \varphi(\overline{x})$, then the formula $\psi(w, \overline{x})$ is satisfiable in $\mathfrak{M}$, and consequently, by $m+1$-elementarity, there is an element x_0 in X such that $\mathfrak{M} \models \psi(x_0, \overline{x})$. Again using the inductive hypothesis, we obtain $\mathfrak{X} \models \psi(x_0, \overline{x})$; consequently, $\mathfrak{X} \models \varphi(\overline{x})$. □

COROLLARY. *A substructure $\mathfrak{X}$ is an elementary substructure of $\mathfrak{M}$ if and only if X is m-elementary for any natural number m.*

In the sequel, instead of saying that $\mathfrak{X}$ is an elementary substructure, we shall often say that X is an elementary substructure.

We define a *envelope of a set X in $\mathfrak{M}$ relative to* a $\varnothing$-definable set A to be a maximal set $\mathrm{en}_A(X)$ such that $\mathrm{en}_A(X)$ and A are independent over X.

REMARK. en_A should not be considered a single-valued operator, associating to a set X its envelope, since there may be more than one envelope.

THEOREM 4.2. *Let A be a strongly minimal $\varnothing$-definable set in an uncountably categorical structure $\mathfrak{M}$. Then for any set X*

(a) $\mathrm{en}_A(X) \supseteq \mathrm{acl}(X)$, $\mathrm{en}_A(X) \cap A = A[X]$;

(b) $\mathrm{en}_A(\mathrm{en}_A(X)) = \mathrm{en}_A(X)$;

(c) *if $A[X]$ is infinite, then $\mathrm{en}_A(X)$ is a prime elementary subsystem of $\mathfrak{M}$ over X;*

(d) *for any natural number m, there exists a natural number s_m such that, if $\|A[X]\| \geq s_m$, then $\mathrm{en}_A(X)$ is m-elementary;*

(e) *if $\mathfrak{M}$ is totally categorical and X is finite, then $\mathrm{en}_A(X)$ is finite.*

PROOF. Throughout the proof we shall write $\mathrm{en}(X)$ instead of $\mathrm{en}_A(X)$.

(a) Let $x \in \mathrm{acl}(X)$. We note that x and A are independent over $\mathrm{en}(X) \cup X$, since $R(x, A \cup \mathrm{en}(X) \cup X) = 0 = R(x, \mathrm{en}(X) \cup X)$. Now, using the symmetry of the definition of independence, we see that

$$R(\overline{y}, \{x\} \cup \mathrm{en}(X) \cup X) = R(\overline{y}, \mathrm{en}(X) \subset X)$$

for any tuple $\overline{y}$ from A, and since by definition $R(\overline{y}, \mathrm{en}(X) \cup X) = R(\overline{y}, X)$, it follows that A and $\{x\} \cup \mathrm{en}(X)$ are independent over X, which is possible only if $x \in \mathrm{en}(X)$.

On the other hand, if $z \in A - A[X]$, then $R(z, A \cup X) = 0$, $0 < R(z, X)$, i.e., $z \notin \mathrm{en}(X)$. □

(b) We shall actually prove more.

If $X \subseteq Y \subseteq \mathrm{en}(X)$, then for any envelope $\mathrm{en}(Y)$ there is an envelope $\mathrm{en}'(X)$ such that $\mathrm{en}(Y) \subseteq \mathrm{en}'(X)$. (*)

Then, taking $Y = \mathrm{en}(X)$ in (*), we obtain

$$\mathrm{en}(X) \subseteq \mathrm{en}(\mathrm{en}(X)) \subseteq \mathrm{en}'(X).$$

By maximality, $\mathrm{en}(X) = \mathrm{en}'(X)$ and this proves the statement in (b).

Let us proceed, then, to the proof of $(*)$. Let $\overline{z}$ be an arbitrary tuple from A. Then, since A and $\mathrm{en}(Y)$ are independent over Y, we have

$$R(\overline{z}, \mathrm{en}(Y)) = R(\overline{z}, Y),$$

but on the other hand, since $X \subseteq Y \subseteq \mathrm{en}(X)$ and A and $\mathrm{en}(X)$ are independent over X,

$$R(\overline{z}, Y) = R(\overline{z}, X).$$

Hence A and $\mathrm{en}(Y)$ are independent over X, and so $\mathrm{en}(Y) \subseteq \mathrm{en}'(X)$ for some maximal set $\mathrm{en}'(X)$ which is independent of A over X. □

(c) Let Y be a subset of the structure which is a prime subsystem over X, and suppose that $A[X]$ is infinite. Since Y is atomic over X, it follows that any element of Y satisfies some X-atom; hence, by the strong minimality of Y, it follows that

$$Y \cap A = A[X].$$

If Y' is a prime subsystem over $\mathrm{en}(X)$, then in just the same way

$$Y' \cap A = A[\mathrm{en}(X)] = A[X]$$

(the last equality follows from the already proved part (a)). By primeness, Y is elementarily embeddable in Y' over X, and since under this embedding the definable subset $A \cap Y$ coincides with $A \cap Y'$, we obtain an isomorphism of Y and Y' over X. In other words, we have proved that a prime subsystem Y over $\mathrm{en}(X)$ is also prime over X.

Now let $\overline{y}$ be a tuple of elements of Y,

$$R(\overline{y}, A \cup X) \leq r.$$

Then for some formula $\varphi(\overline{v}, \overline{w}) \in F$, we have

$$(\exists \overline{z} \in A^k)(\varphi(\overline{y}, \overline{z}) \,\&\, R(\varphi(\mathfrak{M}, \overline{z})) \leq r).$$

Since Y is an elementary subsystem over X, this formula is true in Y if and only if it is true in $\mathfrak{M}$. Hence,

$$R(\overline{y}, A \cup X) = R(\overline{y}, A[X] \cup X).$$

On the other hand, since $A[X] \subseteq \mathrm{acl}(X)$,

$$R(\overline{y}, A[X] \cup X) = R(\overline{y}, X);$$

therefore, Y and A are independent over X. Since Y may be assumed prime over $\mathrm{en}(X)$, it follows that $Y \supseteq \mathrm{en}(X)$. By maximality, we have $Y = \mathrm{en}(X)$. □

(d) We first observe that, if there is a natural number s such that for all X $\|A[X]\| \geq s$ implies that $A[X]$ is infinite, then by part (c) of the theorem it follows from $\|A[X]\| \geq s$ that $\mathrm{en}(X)$ is an elementary subsystem of $\mathfrak{M}$, in particular, that $\mathrm{en}(X)$ is $m+1$-elementary for any m. We may thus assume that for any s there exists X such that $A[X]$ is finite and

$\|A[X]\| \geq s$. Hence, by induction on the number of elements in X, it follows that $A[X]$ is finite for any finite $X \subseteq A$. But then, for any countable elementary substructure X of $\mathfrak{M}$, we have $\dim A[X] = \aleph_0$. This means that any two countable elementary substructures of $\mathfrak{M}$ are isomorphic, that is, the theory of $\mathfrak{M}$ is totally categorical. Throughout the rest of this section, therefore, we shall assume that $\mathfrak{M}$ is totally categorical; hence, by the results of Chapter 3, $\mathfrak{M}$ is a module-like structure.

We shall need the following fact.

AUXILIARY LEMMA 4.3. *In a totally categorical structure, for any $\varnothing$-definable set B of rank 1 there exists a natural number s_B such that if $Z \subseteq A$, $\|Z\| \geq s_B$, then for any set Y*

$$B[Z \cup Y] \neq \varnothing \quad \textit{if and only if} \quad B[A \cup Y] \neq \varnothing.$$

PROOF OF THE LEMMA. We first point out that it will suffice to prove the lemma for strongly minimal B. Indeed, by Shelah's finite equivalence relation theorem, $B = B_1 \cup \cdots \cup B_k$, where each B_i is strongly minimal or finite and definable over $\operatorname{acl}(\varnothing)$. Enriching the signature by constants from $\operatorname{acl}(\varnothing)$ does not alter $\operatorname{acl}(X)$ or, consequently, $B[X]$, for any X. We may therefore assume that all the B_i's are $\varnothing$-definable. But if the required numbers s_{B_i} exist for every $i \leq k$, then we can take s_B to be $\max\{s_{B_i} : i \leq k\}$. We also observe that for finite nonempty sets B_i $B_i[\varnothing] \neq \varnothing$, so we may set $s_{B_i} = 0$ in that case.

Assume, then, that B is strongly minimal. Fix an arbitrary Y_0 for which $B[Y_0] = \varnothing$, but $B[A \cup Y_0] \neq \varnothing$. Choose a maximal subset Z_0 of A for which still $B[Z_0 \cup Y_0] = \varnothing$. Obviously, Z_0 is finite, B is an atom over $Z_0 \cup Y_0$, and for every $b \in B - B[Z_0 \cup Y_0]$ there exists $a \in A - A[Z_0 \cup Y_0]$ such that $\operatorname{acl}(\{b\} \cup Z_0 \cup Y_0) = \operatorname{acl}(\{a\} \cup Z_0 \cup Y_0)$, that is, B is strongly 1-coordinatizable over $A_{Z_0 \cup y_0}$.

By the Polynomial Theorem 1.5.5, there exists a polynomial $r_0(x - s_0)$ such that for any finite $X \supseteq Y_0 \cup Z_0$

$$\|B[X]\| = r_0(x - s_0), \quad \text{where } x = \|A[X]\|.$$

Obviously $s_0 = \|A[Z_0 \cup Y_0]\|$, since $\|B[Z_0 \cup Y_0]\| = 0$.

Now let $B[Y_1] = \varnothing, B[A \cup Y_1] \neq \varnothing$, and let Z_1 be a maximal subset of A for which $B[Z_1 \cup Y_1] = \varnothing$ and

$$\|B[X]\| = r_1(x - s_1) \quad \text{for } X \supseteq Z_1 \cup Y_1 \text{ and } x = \|A[X]\|.$$

For any X that contains $Y_0 \cup Y_1 \cup Z_0 \cup Z_1$ we have

$$r_0(x - s_0) = \|B[X]\| = r_1(x - s_1),$$

and hence $r_0 = r_1$, $s_0 = s_1$. Set $s_B = s_0 + 1$. Now, if $Z \subseteq A$, $\|Z\| \geq s_B$, then for any Y $B[A \cup Y] \neq \varnothing$ implies $B[Z \cup Y] \neq \varnothing$. □

We now continue the proof of part (d) of Theorem 4.2. The proof will make use of the Ladder Theorems 0.2 and 0.3. Using induction on the length of the ladder $A = M_0, M_1, \dots, M_k$, we will prove that for any m there is a natural number $s_{k,m}$ such that, if $\|A[X]\| \geq s_{k,m}$, then $\mathrm{en}(X)$ is m-elementary relative to M_k. This will imply that $\mathrm{en}(X)$ is m-elementary.

We begin the induction with $M_{-1} = \varnothing$. The assertion is obvious for this set. Let $k \geq 0$. Then for every $U \subseteq M$, $b \in M_k$, there exist a finite set $W^{b,U} \subseteq M_{k-1}$ and a $U \cup W^{b,U}$-definable A-atom B containing b, of rank 0 or 1. By countable categoricity, we may assume that $\|W^{b,U}\| \leq n_U$ for all $b \in M_k$. Given an arbitrary m-element set U, let B be $W^{b,U} \cup U$-definable, $b \in B \subseteq M_k$ and B an A-atom.

We then introduce $U \cup W^{b,U}$ into the signature and, using the Auxiliary Lemma, find s_B (depending on the type of $U \cup W^{b,U}$). By countable categoricity, there exists a natural number

$$t_m = \max\{s_B \colon \|U\| \leq m,\ b \in B,\ \|W^{b,U}\| \leq n_u\}.$$

Using the existence of $s_{k-1,n+m}$, we define

$$s_{k,m} = \max\{t_m,\ s_{k-1,n+m}\}.$$

We shall show that this number satisfies the required conditions. Let $\|A[X]\| \geq s_{k,m}$. We must show, given an arbitrary m-element subset $U \subseteq \mathrm{en}(x)$ and a U-definable subset $C \subseteq M_k$, that $C \cap \mathrm{en}(X) \neq \varnothing$.

Let $c \in C$. Since $\|A[X]\| \geq s_{k-1,n+m}$, it follows that $\mathrm{en}(X)$ is $m+n$-elementary relative to M_{k-1}. Consequently, there exists $W \subseteq \mathrm{en}(X)$ such that

$$\mathrm{tp}(\langle u_1, \dots, u_m, w_1, \dots, w_n\rangle) = \mathrm{tp}(\langle u_1, \dots, u_m, w_1^{c,U}, \dots, w_n^{c,U}\rangle),$$

where $\langle u_1, \dots, u_m\rangle$, $\langle w_1, \dots, w_n\rangle$, $\langle w_1^{c,U}, \dots, w_n^{c,U}\rangle$ are orderings of U, W, and $W^{c,U}$, respectively. Therefore, for some $b \in C$

$$W = W^{b,U}, \qquad b \in B \subseteq C,$$

where B is a $U \cup W$-definable A-atom. Since

$$\|A[\mathrm{en}(X)]\| \geq s_{k,m} \geq s_B,$$

it follows that $B[\mathrm{en}(x)] = \varnothing$ implies $B[A \cup \mathrm{en}(X)] = \varnothing$. This means, in particular, that b and A are independent over $\mathrm{en}(X)$, that is, b is in some envelope $\mathrm{en}(\mathrm{en}(X))$. By part (a), $b \in \mathrm{en}(X)$ and consequently $C \cap \mathrm{en}(X) \neq \varnothing$. □

(e) Suppose the contrary. Then, using the fact that the set of complete 1-types over $X \cup \{y_1, \dots, y_k\}$ is finite for any $y_1, \dots, y_k$, we can construct an infinite sequence $\{y_i \colon i \in \omega\}$ in $\mathrm{en}(X)$ such that

$$\mathrm{tp}(y_{i+1}, \{y_1, \dots, y_i\} \cup X) \supsetneq \mathrm{tp}(y_i, \{y_1, \dots, y_{i-1}\} \cup X).$$

Hence, in particular, it follows that $\mathrm{tp}(y_{i+1}, \{y_1, \dots, y_i\} \cup X)$ for any i is realized by any y_j with $j \geq i+1$. From the fact that $\{y_i \colon i \in \omega\} \subseteq$

$\mathrm{en}(X)$ and $\mathrm{en}(X)$ and A are independent over X it follows that for $Y \subseteq \{y_1, \ldots, y_i\}$, $j \geq i+1$,

$$R(y_j, Y \cup A \cup X) = R(y_j, Y \cup X).$$

Since a type over $Y \cup X$ has only finitely many complete extensions over $Y \cup A \cup X$ of the same rank, then the totality of all types $\mathrm{tp}(y_j, Y \cup A \cup X)$, where j ranges over an infinite set, must contain infinitely many equal members. We may therefore extract from $\{y_i : i \in \omega\}$ an infinite subsequence $\{y_{i_k} : k \in \omega\}$ such that

$$\mathrm{tp}(y_{i_{k+1}}, \{y_{i_1}, \ldots, y_{i_k}\} \cup A \cup X) \supseteq \mathrm{tp}(y_{i_k}, \{y_{i_1}, \ldots, y_{i_{k-1}}\} \cup A \cup X).$$

Then for some natural number l and pair $\langle r, d \rangle$ of natural numbers we have

$$\mathrm{RD}(y_{i_j}, \{y_{i_1}, \ldots, y_{i_k}\} \cup A \cup X) = \langle r, d \rangle,$$

if $j > k \geq l$. This implies that the set $\{y_i : j \in \omega, j > l\}$ is indiscernible over $\{y_{i_1}, \ldots, y_{i_l}\} \cup A \cup X$, contrary to the assumption that $\mathfrak{M}$ is uncountably categorical, by 1.2.4. □

COROLLARY 4.4 (finite submodel property). *For any totally categorical structure $\mathfrak{M}$ and any sentence φ which is true in $\mathfrak{M}$, there exists a finite substructure $\mathfrak{L}$ of $\mathfrak{M}$ such that $\mathfrak{L} \models \varphi$.*

PROOF. Let A be a strongly minimal definable subset. Enriching $\mathfrak{M}$ by a finite set of constants, we may assume that A is $\varnothing$-definable. Choosing a finite subset $X \subseteq A$ such that $\|X\| \geq s_k$, where k is the number of quantifiers in φ, we define $L = \mathrm{en}_A(X)$. It follows from Theorem 4.2, parts (d) and (e), and Proposition 4.1 that $\mathfrak{L} \models \varphi$. □

COROLLARY 4.5. *A totally categorical finitely axiomatizable theory is not complete.*

§5. Another version of the proof of the finite submodel property for totally categorical structures

The solution to the finite axiomatizability problem for totally categorical theories follows from the finite submodel property, which was proved in the previous section. The proof of this property relies on the Ladder Theorem for module-like structures. Finally, the proof of the fact that totally categorical structures are module-like or of discrete type constitutes the content of the most difficult part of this dissertation—Chapter 3. In this section we present a proof of the finite submodel property which uses, instead of the main result of Chapter 3, certain number-theoretical arguments, specifically a result of Siegel dating back to 1921.

FACT 5.1 [72]. *If $f(x)$ is a polynomial in one variable with rational coefficients, which takes values of $p_1^{y_1} \cdots p_k^{y_k}$ for fixed $p_1, \dots, p_k$ and suitable natural numbers $y_1, \dots, y_k$ on an infinite set of integer values x, then*

$$f(x) = r(x - s)^d$$

for certain rational numbers r and s and a natural number d.

The Ladder Theorem for module-like structures was used in §4 only in the proof of part (d), and here that theorem may be replaced by the general version of the Ladder Theorem, provided we prove the following analog of the Auxiliary Lemma 4.3.

PROPOSITION 5.2. *In a totally categorical structure, for any $\varnothing$-definable A-atom B and strongly minimal $\varnothing$-definable set A there exists a natural number s_B such that, if $Z \subseteq A$, $\|Z\| \geq s_B$, then for any set Y*

$$B[A \cup Y] \neq \varnothing \quad \textit{if and only if} \quad B[Z \cup Y] \neq \varnothing.$$

The proof of Proposition 5.2 that will be presented below uses the above theorem of Siegel, the results of the second section of this chapter and finally, the Baur-Cherlin-Macintyre Theorem according to which groups definable in totally categorical structures are almost abelian. This theorem may be derived as a direct corollary of Theorems 2.4.1 and 3.0.2, but since the aim of this section is to demonstrate a way of avoiding Theorem 3.0.2, our proof of Proposition 5.2 will implicitly employ the original proof of Baur-Cherlin-Macintyre [40]. We observe, incidentally, that a far simpler fact is actually sufficient for our needs: Groups definable in a totally categorical structure are almost solvable [9], [49].

We now turn to the proof of Proposition 5.2, which will be divided into a series of lemmas.

We first show that it is legitimate to assume that B admits no A-defined equivalence relation with infinite classes.

LEMMA 5.3. *If Proposition 5.2 is valid on the assumption that there exist no A-defined proper equivalence relations on B with infinite classes, then it is valid for any B.*

PROOF. By induction on $\mathrm{RD}(B)$. If E is an A-defined equivalence relation and $\mathrm{RD}(bE) \geq \langle 1, 1 \rangle$, then B/E is an A-defined A-atom and $R(B/E) < R(B)$. By the inductive hypothesis, there exist natural numbers $s_{B/E}$ and s_{bE} for B/E and bE, respectively, where b is an arbitrary element of B. Denote a typical element of B/E by $bE = \hat{b}$ and note that the subset $bE \subseteq B$ is $\hat{b}$-defined. Put $s_B = s_{B/E} + s_{bE}$. Then

$$Z = Z_1 \cup Z_2 \subseteq A, \qquad \|Z_1\| \geq s_{B/E}, \qquad \|Z_2\| \geq s_{BE},$$

and we have

$$\begin{aligned} B[A \cup Y] \neq \varnothing &\Rightarrow (\exists b)(\hat{b} \in (B/E)[A \cup Y] \,\&\, bE[A \cup Y \cup \{\hat{b}\}] \neq \varnothing) \\ &\Rightarrow (\exists b)(\hat{b} \in (B/E)[Z_1 \cup Y] \,\&\, bE[Z_2 \cup Y \cup \{\hat{b}\}] \neq \varnothing) \\ &\Rightarrow B[Z \cup Y] \neq \varnothing. \quad \square \end{aligned}$$

Hence it follows, in particular, that the $\varnothing$-definable equivalence relation $E_{B,A}$ introduced in §2 may be assumed to have finite classes, i.e., $bE_{B,A} \subseteq \mathrm{acl}(b)$ for any $b \in B$. Let E be the equivalence relation defined by the condition

$$xEy \Leftrightarrow x \in \mathrm{acl}(y).$$

Obviously, the assertion of Proposition 5.2 is true for B if and only if it is true for B/E, so we may assume that E, and hence also $E_{B,A}$, are trivial, so that $\mathrm{rd}(B, A) = \langle 0, 1\rangle$. Hence it follows that the group $\mathrm{Gal}(B, A)$ is definable together with its action.

LEMMA 5.4. *If Proposition 5.2 is valid on the assumption that* $\mathrm{Gal}(B, A)$ *is a definable abelian group without proper definable infinite subgroups, then Proposition 5.2 is valid for any* B.

PROOF. As observed previously, we may assume that $\mathrm{Gal}(B, A)$ is definable. We use the Baur-Cherlin-Macintyre Theorem [40], which states that groups definable in totally categorical structures are almost abelian, and find a normal definable abelian subgroup G such that $\mathrm{Gal}(B, A)/H$ is finite. Then by Proposition 2.9 there is an A-definable equivalence relation E_H^* on B^* such that B^*/E_H^* is finite, and $\mathrm{Gal}(B, A \cup (B^*/E_H^*)) = H$. Therefore, introducing the finite set of constants B^*/E_H^* into the signature, we get the abelian group H as Galois group. The rest follows from Lemma 5.3 and Propositions 2.7, 2.8. $\square$

REMARK. We could have used here a considerably simpler result than the Baur-Cherlin-Macintyre Theorem: Groups definable in totally categorical structures are almost solvable.

We now continue the proof of Proposition 5.2, assuming that B is an A-atom, the group $\mathrm{Gal}(B, A)$ is $\varnothing$-definable with its action, and $\mathrm{Gal}(B, A)$ is abelian and has no proper infinite definable subgroups. Henceforth we denote $\mathrm{Gal}(B, A) = G$. It follows from our assumptions that G is connected, i.e., $D(G) = 1$.

It follows from Propositions 2.3, 2.5 that

$$G \subseteq \mathrm{acl}(A).$$

By enriching the signature with a finite set of constants from A, we may assume that G is strictly coordinatizable over A. Consequently, by the Polynomial Theorem, there exists a polynomial $f(x)$ with rational coefficients such that for all finite X

(1) if $\|A[X]\| = x$, then $\|G[X]\| = f(x)$.

Note, moreover, that G is an abelian group of prime exponent p. Obviously $G[X]$ is a subgroup of G, and so

(2) $f(x) = \|G[X]\| = p^y$.

Hence, by Siegel's Theorem,

(3) $f(x) = r(x - w)^m$,

where $m = R(G)$ by the Polynomial Theorem.

Fix a basis $\{a_1, \dots, a_i, \dots\}$ of the strongly minimal set A and denote

$$u_i = \|A[a_1, \dots, a_i]\|, \qquad \tilde{u}_i = u_i - w.$$

Then, in view of (2), and (3), we see that for $i \geq 0$

$$\tilde{u}_{i+1}/\tilde{u}_i = (f(u_{i+1})/f(u_i))^{1/m} = p^{z_i/m},$$

where z_i is some positive integer. Since the fraction calculated here is rational, it follows that z_i/m is a positive integer, i.e.,

(4) $\tilde{u}_{i+1}/\tilde{u}_i = p^{k_i}$, where k_i is a positive integer.

Let $C_1, \dots, C_m$ be arbitrary strongly minimal subsets of G and $f_i(x)$ polynomials such that for X such that C_i is X-definable

$$x = \|A[X]\| \Rightarrow \|C_i[X]\| = f_i(x).$$

Since $R(C_i) = 1$,

$$f_i(x) = t_i x - v_i$$

for some rational t_i, v_i. To simplify matters, we shall assume that the C_i are $\varnothing$-definable sets.

We claim that

(5) $r^{-1} \cdot t_1 \cdots t_m$ is an integer.

Indeed, consider the map

$$\Pi \colon C_1 \times \cdots \times C_m \to G$$

with domain of definition

$$\operatorname{Dom}\Pi = \{\langle g_1, \dots, g_m\rangle \in C_1 \times \cdots \times C_m \colon g_1, \dots, g_m \text{ are algebraically independent}\},$$

defined by the condition

$$\Pi(g_1, \dots, g_m) = g_1 \cdots g_m.$$

Let $C^{(k)}$ denote $C_1 \cdots C_k$ $(k \leq m)$. Obviously, $R(C^{(k)}) \leq k$. We claim that if $k \leq m$, then $R(C^{(k)}) = k$. Indeed, otherwise, if $R(C^{(k)}) = R(C^{(k-1)}) = k - 1$, there exists $g_0 \in C_k$ such that for infinitely many g in C_k

$$R(g \cdot C^{(k-1)} \cap g_0 \cdot C^{(k-1)}) = k - 1,$$

and hence,

$$g \cdot q = g_0 \cdot q, \qquad R(q) = k - 1$$

for some type q containing $C^{(k-1)}$ and infinitely many numbers g in C_k. This implies that the stabilizer $\mathrm{St}_G(q)$ of q is infinite (Chapter 1, §4), but then $\mathrm{St}_G(q)$ is an infinite definable subgroup of G, $R(\mathrm{St}_G(q)) \le R(q) = k-1 < R(G)$.

Thus $R(C^{(m)}) = m$. It is also easy to see that since $D(G) = 1$, $G \sqsubset C^{(m)} \sqsubset \text{Range } \Pi$, that is, $R(\text{Range } \Pi) = m$ and therefore the preimage $\Pi^{-1}(g)$ of each point g is finite.

By construction, Range Π is an atom, and therefore the number of elements in $\Pi^{-1}(g)$ is independent of $g \in \text{Range } \Pi$. Denote this number by 1. It is easily seen by a direct calculation that

$$\text{Range } \Pi[X] = l^{-1} \cdot (t_1 x - v_1') \cdots (t_m x - v_m') = \overline{f}(x),$$

where $v_i' = v_i + \|C_i[g_1, \dots, g_{i-1}]\|$, $i \le m$, $\langle g_1, \dots, g_{i-1}\rangle \in C_1 \times \cdots \times C_{i-1}$, $g_1, \dots, g_{i-1}$ are algebraically independent. Since $G \sqsubset \text{Range } \Pi$, it follows that $R(G - \text{Range } \Pi) < m$ and therefore the leading coefficients of the polynomials f and $\overline{f}$ are equal, i.e.,

$$l^{-1} \cdot t_1 \cdots t_m = r \quad \text{or} \quad r^{-1} \cdot t_1 \cdots t_m = l,$$

which proves (5).

As a corollary, we obtain

(5′) If $C \subseteq G$, $R(C) = 1$, $p_C(x) = \|C[X]\| = tx - v$, then $r^{-1} \cdot t^m$ is an integer.

Indeed, C may be expressed as a union of finitely many strongly minimal sets $C = \bigcup_{i=1}^n C_i$; then $t = \sum_{i=1}^n t_i$, where t_i is the coefficient of x in the polynomial $p_{C_i}(x)$. Property (5′) now follows from (5) by direct calculation of $r^{-1} \cdot t^m$.

We now study the exponents k_i in (4). To simplify matters, let us assume that $i = 0$. Then

$$\|G[a_1]/G[\varnothing]\| = \frac{f(u_1)}{f(u_0)} = p^{k_0 \cdot m},$$

$$\|G[a_1, a_2]/G[a_2]\| = \frac{f(u_2)}{f(u_1)} = p^{k_1 \cdot m},$$

and since $G[a_1] \cap G[a_2] = G[\varnothing]$

$$p^{(k_1+k_0)\cdot m} = \|G[a_1, a_2]/G[\varnothing]\| \ge \|G[a_1] \cdot G[a_2]/G[\varnothing]\| = p^{2k_0 \cdot m}.$$

This inequality implies that $k_1 \ge k_0$, i.e.,

(6) $k_{i+1} \ge k_i$,

and equality $k_{i+1} = k_i$ implies

(7) $G[a_1, \dots, a_i, a_{i+1}, a_{i+2}] = G[a_1, \dots, a_i, a_{i+1}] \cdot G[a_1, \dots, a_i, a_{i+2}]$.

We shall show now that (7) implies $R(G) = 1$. We shall again assume that $i = 0$.

Let $H = \bigcup\{G[a] \colon a \in A\}$.

Clearly, $R(H) = 1$ and $h \in H$ if and only if $h^{-1} \in H$. Since $\tilde{u}_{i+1}/\tilde{u}_i = p^{k_i} \geq 2$ for all i, it follows that

$$u_2 - u_0 > 2(u_1 - u_0).$$

This means that there exists $a \in A[a_1, a_2] - (A[a_1] \cup A[a_2])$. Therefore, there exists

$$h \in H \cap (G[a_1, a_2] - (G[a_1] \cup G[a_2])).$$

By (7), $h = h_1 \cdot h_2$ for some $h_1 \in G[a_1] \subseteq H$, $h_2 \in G[a_2] \subseteq H$. Therefore, $h \cdot h_2^{-1} \in H$ for some $h_2^{-1} \in H$. But $h_2 \notin \operatorname{acl}(h)$, $h \notin \operatorname{acl}(\varnothing)$, whence it follows that $h \cdot H \cap H$ is infinite for infinitely many h in H. It is now easily shown that the stabilizer $\operatorname{St}_G(q)$ of some type q containing H is infinite. Since $0 < R(\operatorname{St}_G(q)) \leq R(q) \leq R(H) \leq 1$, while at the same time G does not contain proper infinite definable subgroups, it follows that $R(G) = 1$.

We shall now show that we may assume

(8) $k_{i+1} > k_i$ for all $i \geq 0$.

Indeed, otherwise $R(G) = 1$, but since B is an orbit of G, it would follow that also $R(B) = 1$, and in that case Proposition 5.2 coincides with Auxiliary Lemma 4.3.

Now, assuming that (8) is true, we shall prove Proposition 5.2. Put $s_B = u_1$. Suppose, contrary to what we are trying to prove, that $\|Z\| \geq s_B$, $Z \subseteq A$, $B[Z \cup Y] = \varnothing$, but $B[A \cup Y] \neq \varnothing$. This last relation, by Proposition 2.8, implies that $B \subseteq \operatorname{acl}(A \cup Y)$. We may assume that Z is the maximal set of all those for which $B[Z \cup Y] = \varnothing$, i.e., $B[Z \cup Y \cup \{a\}] \neq \varnothing$ for any $a \in A - A[Z]$.

Denote

$$Z \cup Y = X_0 \quad \text{and} \quad B_0 = \bigcup \{B[X_0 \cup \{a\}] : a \in A\} - B[X_0].$$

$u'_i = \|A[X_0 \cup \{a'_1, \ldots, a'_i\}]\|$, where $a'_1, \ldots, a'_i \in A$ are algebraically independent over X_0, $\tilde{u}'_i = u'_i - w$. It is clear that B_0 is X_0-definable, $R(B_0) = 1$.

Consider a finite set $X \supseteq X_0$, $x = \|A[X]\|$. The set $B_0[X]$ can be expressed as a union of disjoint subsets $B[X_0 \cup \{a\}] - B[X_0]$ for $a \in A[X] - A[X_0]$, the number of subsets in the union being equal to the number of subsets of the form $A[X_0 \cup \{a\}] - A[X_0]$ with $a \in A[X] - A[X_0]$. The latter is obviously equal to $\frac{x - u'_0}{u'_1 - u'_0}$. Thus, for any $X \supseteq X_0$, if $x = \|A[X]\|$, then

$$p_{B_0}(x) = \|B_0[X]\| = t(x - u'_0), \quad \text{where } t = \frac{\|B_0[X_0 \cup \{a\}]\|}{u'_1 - u'_0}, \quad a \in A - A[X_0].$$

Choosing $b \in B_0[X_0 \cup \{a\}]$, consider the map $\alpha_b \colon G \to B$, where $\alpha_b(g) = b \blacksquare g$ for $g \in G$. The map α_b establishes a one-to-one correspondence between

G and B;

$G[X_0 \cup \{a\}]$ and $B[X_0 \cup \{a\}]$;
$G[X]$ and $B[X]$;
C and B_0, where $C = \alpha_b^{-1}(B_0)$.

In particular, $\|B[X_0 \cup \{a\}]\| = \|G[X_0 \cup \{a\}\| = p^{l_1}$ for some natural number l_1, and

$$p_{B_0}(x) = p_C(x) = t \cdot (x - u_0'), \quad \text{where } t = p^{l_1}(u_1' - u_0')^{-1} = p^{l_1}(\tilde{u}_1' - \hat{u}_0')^{-1}.$$

Since $\tilde{u}_1'/\tilde{u}_0' = p^k$ for some positive natural k, we have $t = p^{l_1}(p^k - 1)^{-1} \cdot \tilde{u}_0'^{-1}$. We now calculate

$$\begin{aligned} r^{-1} \cdot t^m &= r^{-1} \cdot \tilde{u}_0'^{-m} \cdot p^{l_1 \cdot m} \cdot (p^k - 1)^m = f(u_0')^{-1} \cdot p^{l_1 \cdot m} \cdot (p^k - 1)^{-m} \\ &= p^{l_1 \cdot m - l_0}(p^k - 1)^{-m}, \end{aligned}$$

where $p^{l_0} = f(u_0')$. In view of $(5')$, this number is an integer, which is possible only if $p = 2$, $k = 1$.

Obviously, $\tilde{u}_1' = \tilde{u}_{i_1}$, $\tilde{u}_0' = \tilde{u}_{i_0}$ for some $i_1 > i_0 \geq 1$. Therefore,

$$\sum_{i=i_0}^{i_1} k_i = k = 1,$$

which is possible only if $i_1 = i_0 + 1$, $k_{i_0} = 1$. By (6), $k_{i_0} \geq k_1 \geq k_0$; therefore $k_1 = k_0$, contrary to (8). □

Bibliography

1. A. I. Abakumov, E. A. Palyutin, M. A. Taitslin, and Yu. E. Shishimarev, *Categorical quasivarieties*, Algebra i Logika **2** (1972), no. 1, 3–38; English transl. in Algebra and Logic **2** (1972).
2. M. Aigner, *Combinatorial theory*, Springer-Verlag, Berlin and New York, 1979.
3. M. M. Erimbetov, *Categoricity in power and nonbicardinal formulas of finite rank*, Algebra i Logika **13** (1974), no. 5, 493–500; English transl. in Algebra and Logic **13** (1974).
4. ———, *On complete theories with* 1*-cardinal formulas*, Algebra i Logika **14** (1975), 245–257; English transl. in Algebra and Logic **14** (1975).
5. B. I. Zilber, *On the transcendence rank of formulas of* $\aleph_1$*-categorical theories*, Mat. Zametki **15** (1974), no. 2, 321–329; English transl. in Math. Notes **15** (1974).
6. ———, *Rings whose theories are* $\aleph_1$*-categorical*, Algebra i Logika **13** (1974), no. 2, 168–187; English transl. in Algebra and Logic **13** (1974).
7. ———, *Groups and rings whose theories are categorical*, Fund. Math. **95** (1977), no. 2, 173–188.
8. ———, *Combinatorial problems in categoricity theory*, Abstracts of Reports, 6th Kazakhstan Interuniversity Conference on Mathematics and Mechanics, Part 1, 1977, p. 126. (Russian)
9. ———, *Structure of models of categorical theories and the finite axiomatizability problem*, Preprint No. 2800-77, deposited at VINITI, 1977. (Russian)
10. ———, *Definability of algebraic systems and categoricity theory*, Fifth All-Union Conference on Mathematical Logic (Novosibirsk, 1979), Inst. Math., Siberian Branch, 1979, p. 55. (Russian)
11. ———, *Strongly minimal countably categorical theories*, Sibirsk. Mat. Zh. **21** (1980), no. 2, 98–112; English transl. in Siberian Math. J. **21** (1980).
12. ———, *Solution of the problem of finite axiomatizability for theories that are categorical in all infinite powers*, Theory of Models and its Applications, Kazakh. Gos. Univ., Alma-Ata, 1980, pp. 47–60. (Russian)
13. ———, *Totally categorical structures and combinatorial geometries*, Dokl. Akad. Nauk SSSR **259** (1981), no. 5, 1039–1041; English transl. in Soviet Math. Dokl. **24** (1981).
14. ———, *On the finite axiomatizability problem of theories categorical in all infinite powers*, Investigations in Theoretical Programming, Kazakh. Gos. Univ., Alma-Ata, 1981, pp. 69–74. (Russian)
15. ———, *On abstract dependence systems*, XXVIth All-Union Algebraic Conference, Part 1 (Leningrad, 1981), LOMI, Leningrad, 1981, p. 56. (Russian)
16. ———, *Strongly minimal countably categorical theories* II, Sibirsk. Mat. Zh. **25** (1984), no. 3, 71–88; English transl. in Siberian Math. J. **25** (1984).
17. ———, *Strongly minimal countably categorical theories* III, Sibirsk. Mat. Zh. **25** (1984), no. 4, 63–77; English transl. in Siberian Math. J. **25** (1984).
18. ———, *On definability of a pseudoplane in rings and groups*, Fifth All-Union Symposium on Theory of Rings, Algebras and Modules (Novosibirsk, 1982), Inst. Math., Siberian Branch, 1982, p. 59. (Russian)

19. ———, *Description of certain transitive groups*, XVIIth All-Union Algebraic Conference, Part 1 (Minsk, 1983), Inst. Math. AN BSSR, Minsk, 1983, p. 79. (Russian)
20. ———, *On universally axiomatizable uncountably categorical theories*, VIIth All-Union Conference on Mathematical Logic (Novosibirsk, 1984), Inst. Math., SO AN SSSR, 1984. (Russian)
21. H. J. Keisler and C. C. Chang, *Model theory*, North-Holland, Amsterdam, 1977.
22. E. A. Palyutin, *Models with countably categorical universal theories*, Algebra i Logika **10** (1971), no. 1, 23–32; English transl. in Algebra and Logic **10** (1971).
23. ———, *Categorical quasivarieties of arbitrary signature*, Sibirsk. Mat. Zh. **14** (1973), no. 6, 1285–1303; English transl. in Siberian Math. J. **14** (1973).
24. ———, *Description of categorical quasivarieties*, Algebra i Logika **14** (1975), no. 2, 145–185; English transl. in Algebra and Logic **14** (1975).
25. ———, *On categorical positive Horn theories*, Algebra i Logika **18** (1979), no. 1, 47–72; English transl. in Algebra and Logic **18** (1979).
26. ———, *Categorical Horn classes*. I, Algebra i Logika **19** (1980), no. 5, 582–614; English transl. in Algebra and Logic **19** (1980).
27. ———, *Description of categorical positive Horn classes*, Algebra i Logika **19** (1980), no. 6, 683–700; English transl. in Algebra and Logic **19** (1980).
28. M. G. Peretyat′kin, *An example of an $\aleph_1$-categorical complete finitely axiomatizable theory*, Algebra i Logika **19** (1980), no. 3, 314–347; English transl. in Algebra and Logic **19** (1980).
29. G. E. Sacks, *Saturated model theory*, Benjamin, Reading, MA, 1972.
30. M. A. Taitslin, *A non almost strongly minimal theory which is categorical in infinite powers*, 2nd All-Union Conference on Mathematical Logic (Moscow, 1972), Inst. Math. AN SSSR, Moscow, 1972, pp. 47–48. (Russian)
31. S. A. Chikhachev, *$\aleph_1$-categorical commutative rings*, Sibirsk. Mat. Zh. **18** (1977), no. 4, 908–914; English transl. in Siberian Math. J. **18** (1977).
32. ———, *Categorical universal classes of semigroups*, Sibirsk. Mat. Zh. **16** (1975), no. 1, 149–157; English transl. in Siberian Math. J. **16** (1975).
33. Yu. E. Shishimarev, *On categorical theories of one function*, Mat. Zametki **11** (1972), no. 1, 89–98; English transl. in Math. Notes **11** (1972).
34. G. Ahlbrandt, *On totally categorical theories*, Abstracts Amer. Math. Soc. **5** (1984), no. 2, 158.
35. J. T. Baldwin, *Almost strongly minimal theories*. I, J. Symbolic Logic **37** (1972), 487–493.
36. ———, *Almost strongly minimal theories*. II, J. Symbolic Logic **37** (1972), 657–660.
37. ———, *α_T is finite for $\aleph_1$-categorical T*, Trans. Amer. Math. Soc. **181** (1973), 37–51.
38. J. T. Baldwin and A. H. Lachlan, *On strongly minimal sets*, J. Symbolic Logic **36** (1971), 79–96.
39. ———, *On universal Horn classes categorical in some infinite power*, Algebra Universalis **3** (1973), 98–111.
40. W. Baur, G. Cherlin, and A. Macintyre, *Totally categorical groups and rings*, J. Algebra **57** (1979), 407–440.
41. S. Buechler, *One theorem of Zilber's on strongly minimal sets*, J. Symbolic Logic **50** (1985), no. 4, 1054–1061.
42. ———, *Coordinatization in superstable theories*. I, *Stationary types*, Trans. Amer. Math. Soc. **288** (1985), 101–114.
43. G. Cherlin, *Groups of small Morley rank*, Ann. Math. Logic **17** (1979), 1–28.
44. G. Cherlin, L. Harrington, and A. H. Lachlan, *$\aleph_0$-categorical, $\aleph_0$-stable theories*, Ann. Pure Appl. Logic **28** (1985), 103–135.
45. H. Crapo and G.-C. Rota, *On the foundation of combinatorial theory: Combinatorial geometries*, MIT Press, Cambridge, MA and London, 1970.
46. P. Dembowski, *Finite geometries*, Springer-Verlag, Berlin and New York, 1968.
47. J. Doyen and X. Hubaut, *Finite regular locally projective spaces*, Math. Z. **119** (1971), no. 1, 83–88.
48. U. Felgner, *$\aleph_1$-Kategorische theorien nicht-kommutativer Ringe*, Fund. Math. **82** (1974), 331–346.

49. ———, *$\aleph_0$-categorical stable groups*, Math. Z. **160** (1978), 27–49.
50. S. Givant, *Universal Horn classes categorical or free in power*, Ann. Math. Logic **15** (1978), 1–53.
51. ———, *A representation theorem for universal Horn classes categorical in power*, Ann. Math. Logic **17** (1979), 91–96.
52. A. H. Lachlan, *Two conjectures regarding the stability of ω-categorical theories*, Fund. Math. **81** (1974), 133–145.
53. ———, *Theories with a finite number of models in an uncountable power are categorical*, Pacific J. Math. **61** (1975), 465–481.
54. D. Lascar, *Ranks and definability in superstable theories*, Israel J. Math. **23** (1976), 53–87.
55. P. Lindstrom, *On model completeness*, Theoria **30** (1964), 183–196.
56. J. Los, *On the categoricity in power of elementary deductive systems and some related problems*, Colloq. Math. **3** (1964), 58–62.
57. J. Loveys, *The uniqueness of envelopes in $\aleph_0$-categorical, $\aleph_0$-stable structures*, J. Symbolic Logic **49** (1984), 1171–1184.
58. A. Macintyre, *On $\aleph_1$-categorical theories of Abelian groups*, Fund. Math. **70** (1971), 253–270.
59. ———, *On $\aleph_1$-categorical theories of fields*, Fund. Math. **71** (1971), 1–25.
60. J. A. Makowsky, *On some conjectures connected with complete sentences*, Fund. Math. **81** (1974), 193–202.
61. E. Marczewski, *Independence in some abstract algebras*, Bull. Acad. Polon. Sci. Ser. Sci. Math. Astr. Phys. **7** (1959), 611–616.
62. W. Marsh, *On $\aleph_1$- but not $\aleph_0$-categorical theories*, Ph.D. Thesis, University of Dartmouth, 1966.
63. J. Mazoyer, *Sur les théories catégoriques finiment axiomatisables*, C. R. Acad. Sci. Paris. Sér. I Math. **281** (1975), 403–406.
64. D. Morley, *Categoricity in power*, Trans. Amer. Math. Soc. **114** (1965), 514–538.
65. W. Narkiewicz, *Independence in a certain class of abstract algebras*, Fund. Math. **50** (1961), 333–340.
66. E. A. Palyutin, *Completeness and categoricity of quasivarieties*, 6th International Congress of Logic, Methodology and Philosophy of Science (Hanover, 1979), pp. 83–87.
67. ———, *Criterion of complete varieties categoricity*, 6th International Congress of Logic, Methodology and Philosophy of Science (Hanover, 1979), pp. 143–147.
68. A. Pillay, *Stable theories, pseudoplanes and the number of countable models*, Ann. Pure Appl. Logic **43** (1989), no. 2, 147–160.
69. J. Reineke, *Minimale Gruppen*, Z. Math. Logik Grundlag Math. **21** (1975), 357–368.
70. B. I. Rose, *Model theory of alternative rings*, Notre Dame J. Formal Logic **19** (1978), 215–247.
71. ———, *The $\aleph_1$-categoricity of strictly upper triangular matrix rings over algebraically closed fields*, J. Symbolic Logic **43** (1978), 250–259.
72. K. L. Siegel, *Approximation algebraischer Zahlen*, Math. Z. **10** (1921), 173–213.
73. S. Shelah, *Categoricity of uncountable theories*, Proceedings of the Tarski Symposium, Amer. Math. Soc., Providence, RI, 1974, pp. 187–203.
74. ———, *Classification theory and number of nonisomorphic models*, North-Holland, Amsterdam, 1978.
75. K. Urbanik, *A representation theorem for v^*-algebras*, Fund. Math. **53** (1963), 291–317.
76. R. Vaught, *Models of complete theories*, Bull. Amer. Math. Soc. **69** (1963), 299–313.
77. O. Veblen and J. Young, *Projective geometry*, Ginn, Boston, 1946.
78. B. L. Van der Waerden, *Moderne Algebra*, Springer-Verlag, Berlin and New York, 1937.
79. M. Ziegler, *Model theory of modules*, Ann. Pure Appl. Logic **29** (1984), no. 2, 149–213.
80. B. I. Zilber, *Some problems concerning categorical theories*, Bull. Acad. Polon. Sci. Ser. Sci. Math. **29** (1981), 47–49.
81. ———, *Totally categorical theories: Structural properties and the nonfinite axiomatizability*, Model Theory of Algebra and Arithmetic (Proc. Conf., Karpacz, 1979), Lecture Notes in Math., vol. 834, Springer-Verlag, Berlin and New York, 1980, pp. 381–410.

82. ———, *The structure of models of uncountably categorical theories*, Proceedings of the International Congress of Mathematicians (Warsaw, 1983), vol. 1–2, PWN, Warsaw, 1984, pp. 359–368.
83. ———, *Structural properties of models of $\aleph_1$-categorical theories*, Logic, Methodology and Philosophy of Sciences. VII, North-Holland, Amsterdam and New York, 1986, pp. 115–128.
84. A. Pillay, *Stable groups without pseudoplanes*, Abstracts Amer. Math. Soc. **6** (1985), 226.
85. D. Evans, *Homogeneous geometries*, Ph.D. Thesis, Queens College, Oxford, 1985.
86. P. Neumann, *Some primitive permutation groups*, Proc. London Math. Soc. (3) **50** (1985), 265–281.

Subject Index